LONDON MATHEMATICAL SOCIETY LECTURE NOTE SERIE

Managing Editor: Professor J.W.S. Cassels, Department of Pure Mathematics and Mathematical Statistics, 16 Mill Lane, Cambridge CB2 1SB, England

4 Algebraic topology, J.F.ADAMS
5 Commutative algebra, J.T.KNIGHT
8 Integration and harmonic analysis on compact groups, R.E.EDWARDS
11 New developments in topology, G.SEGAL (ed.)
12 Symposium on complex analysis, J.CLUNIE & W.K.HAYMAN (eds.)
13 Combinatorics, T.P.McDONOUGH & V.C.MAVRON (eds.)
15 An introduction to topological groups, P.J.HIGGINS
16 Topics in finite groups, T.M.GAGEN
17 Differential germs and catastrophes, Th.BROCKER & L.LANDER
18 A geometric approach to homology theory, S.BUONCRISTIANO, C.P. ROURKE & B.J.SANDERSON
20 Sheaf theory, B.R.TENNISON
21 Automatic continuity of linear operators, A.M.SINCLAIR
23 Parallelisms of complete designs, P.J.CAMERON
24 The topology of Stiefel manifolds, I.M.JAMES
25 Lie groups and compact groups, J.F.PRICE
26 Transformation groups, C.KOSNIOWSKI (ed)
27 Skew field constructions, P.M.COHN
29 Pontryagin duality and the structure of LCA groups, S.A.MORRIS
30 Interaction models, N.L.BIGGS
31 Continuous crossed products and type III von Neumann algebras, A.VAN DAELE
32 Uniform algebras and Jensen measures, T.W.GAMELIN
34 Representation theory of Lie groups, M.F. ATIYAH et al.
35 Trace ideals and their applications, B.SIMON
36 Homological group theory, C.T.C.WALL (ed.)
37 Partially ordered rings and semi-algebraic geometry, G.W.BRUMFIEL
38 Surveys in combinatorics, B.BOLLOBAS (ed.)
39 Affine sets and affine groups, D.G.NORTHCOTT
40 Introduction to Hp spaces, P.J.KOOSIS
41 Theory and applications of Hopf bifurcation, B.D.HASSARD, N.D.KAZARINOFF & Y-H.WAN
42 Topics in the theory of group presentations, D.L.JOHNSON
43 Graphs, codes and designs, P.J.CAMERON & J.H.VAN LINT
44 Z/2-homotopy theory, M.C.CRABB
45 Recursion theory: its generalisations and applications, F.R.DRAKE & S.S.WAINER (eds.)
46 p-adic analysis: a short course on recent work, N.KOBLITZ
47 Coding the Universe, A.BELLER, R.JENSEN & P.WELCH
48 Low-dimensional topology, R.BROWN & T.L.THICKSTUN (eds.)
49 Finite geometries and designs, P.CAMERON, J.W.P.HIRSCHFELD & D.R.HUGHES (eds)
50 Commutator calculus and groups of homotopy classes, H.J.BAUES
51 Synthetic differential geometry, A.KOCK
52 Combinatorics, H.N.V.TEMPERLEY (ed)
53 Singularity theory, V.I.ARNOLD
54 Markov process and related problems of analysis, E.B.DYNKIN
55 Ordered permutation groups, A.M.W.GLASS
56 Journées arithmétiques, J.V.ARMITAGE (ed)
57 Techniques of geometric topology, R.A.FENN
58 Singularities of smooth functions and maps, J.A.MARTINET
59 Applicable differential geometry, M.CRAMPIN & F.A.E.PIRANI
60 Integrable systems, S.P.NOVIKOV et al
61 The core model, A.DODD

62 Economics for mathematicians, J.W.S.CASSELS
63 Continuous semigroups in Banach algebras, A.M.SINCLAIR
64 Basic concepts of enriched category theory, G.M.KELLY
65 Several complex variables and complex manifolds I, M.J.FIELD
66 Several complex variables and complex manifolds II, M.J.FIELD
67 Classification problems in ergodic theory, W.PARRY & S.TUNCEL
68 Complex algebraic surfaces, A.BEAUVILLE
69 Representation theory, I.M.GELFAND et al.
70 Stochastic differential equations on manifolds, K.D.ELWORTHY
71 Groups - St Andrews 1981, C.M.CAMPBELL & E.F.ROBERTSON (eds.)
72 Commutative algebra: Durham 1981, R.Y.SHARP (ed.)
73 Riemann surfaces: a view towards several complex variables,A.T.HUCKLEBERRY
74 Symmetric designs: an algebraic approach, E.S.LANDER
75 New geometric splittings of classical knots, L.SIEBENMANN & F.BONAHON
76 Linear differential operators, H.O.CORDES
77 Isolated singular points on complete intersections, E.J.N.LOOIJENGA
78 A primer on Riemann surfaces, A.F.BEARDON
79 Probability, statistics and analysis, J.F.C.KINGMAN & G.E.H.REUTER (eds.)
80 Introduction to the representation theory of compact and locally compact groups, A.ROBERT
81 Skew fields, P.K.DRAXL
82 Surveys in combinatorics, E.K.LLOYD (ed.)
83 Homogeneous structures on Riemannian manifolds, F.TRICERRI & L.VANHECKE
84 Finite group algebras and their modules, P.LANDROCK
85 Solitons, P.G.DRAZIN
86 Topological topics, I.M.JAMES (ed.)
87 Surveys in set theory, A.R.D.MATHIAS (ed.)
88 FPF ring theory, C.FAITH & S.PAGE
89 An F-space sampler, N.J.KALTON, N.T.PECK & J.W.ROBERTS
90 Polytopes and symmetry, S.A.ROBERTSON
91 Classgroups of group rings, M.J.TAYLOR
92 Representation of rings over skew fields, A.H.SCHOFIELD
93 Aspects of topology, I.M.JAMES & E.H.KRONHEIMER (eds.)
94 Representations of general linear groups, G.D.JAMES
95 Low-dimensional topology 1982, R.A.FENN (ed.)
96 Diophantine equations over function fields, R.C.MASON
97 Varieties of constructive mathematics, D.S.BRIDGES & F.RICHMAN
98 Localization in Noetherian rings, A.V.JATEGAONKAR
99 Methods of differential geometry in algebraic topology, M.KAROUBI & C.LERUSTE
100 Stopping time techniques for analysts and probabilists, L.EGGHE
101 Groups and geometry, ROGER C.LYNDON

London Mathematical Society Lecture Note Series. 101

Groups and geometry

ROGER C. LYNDON
University of Michigan

Revised from Groupes et Géométrie,
by A. Boidin, A. Fromageot, and R. Lyndon,
Cours 1980-81, Université de Picardie, Amiens

The right of the
University of Cambridge
to print and sell
all manner of books
was granted by
Henry VIII in 1534.
The University has printed
and published continuously
since 1584.

CAMBRIDGE UNIVERSITY PRESS

Cambridge

London New York New Rochelle

Melbourne Sydney

Published by the Press Syndicate of the University of Cambridge
The Pitt Building, Trumpington Street, Cambridge CB2 1RP
32 East 57th Street, New York, NY 10022, USA
10, Stamford Road, Oakleigh, Melbourne 3166, Australia

© Cambridge University Press 1985

First published 1985

Printed in Great Britain at the University Press, Cambridge

Library of Congress catalogue card number: 84-21484

British Library cataloguing in publication data

Lyndon Roger C.
Groups and geometry - (London Mathematical
Society lecture note series, ISSN 0076-0052; 101)
1. Groups, Theory of. 2. Combinatorial analysis
I. Title II. Boidin, A. Groupes et Géométrie
III. Series
512'.22 QA171

ISBN 0 521 31694 4

PREFACE

This book is intended as an introduction, demanding a minimum of background, to some of the central ideas in the theory of groups and in geometry. It grew out of a course, for advanced undergraduates and beginning graduate students, given several times at the University of Michigan and, in 1980-81, at the Université de Picardie. It is assumed that the reader has some acquaintance with the algebra of the complex plane, with analytic geometry, and with the basic concepts of linear algebra. No technical knowledge of geometry is assumed, and no knowledge of group theory, although some exposure to the fundamental ideas of group theory would probably prove helpful.

We exploit the well known close connections between group theory and geometry to develop the two subjects in parallel. Group theory is used to clarify and unify the geometry, while the geometry provides concrete and intuitive examples of groups. This has some influence on our emphasis, which is primarily combinatorial. The groups are mainly infinite groups, often given by generators and relations. The geometry is mainly incidence geometry; apart from linear algebra, we have used analytic and metric methods sparingly. In the interest of intuition we have, with one exception, confined attention to two-dimensional geometry.

Except in connection with projective geometry, we have paid little attention to axioms. We feel that this is no real loss of rigor, since, where intuition is not found sufficient, the reader can always fall back on analytic geometry to verify elementary assertions.

The emphasis has been on geometric spaces and their groups, rather than on theorems concerning special configurations. There are three chapters that lie outside our systematic progress through the study of the classical geometric spaces, and which may be regarded as optional. Each shows the general abstract ideas put to work on special problems. Chapter IV classifies the planar crystallographic groups. Chapter V classifies the regular tessellations of the higher dimensional spheres (and thus the regular solids) and Euclidean spaces, using very little beyond two-dimensional arguments. Chapter X, after developing basic concepts in the theory of Fuchsian groups, is mainly devoted to examples.

Throughout, while emphasizing abstract concepts, we have illustrated these concepts with concrete and intuitive examples, both in the text and in the problems that follow each chapter.

Where appropriate we have tried to guide the reader by references, both to collateral or alternative expositions and to sources for material beyond that presented here. A substantial list of references appears at the end of the book.

The author is grateful to his former students for their interest, criticism, and suggestions. He wants to thank his colleagues A. Boidin and A. Fromageot for invaluable assistance both with his course at Amiens and in the preparation of the resulting notes. He especially thanks Antoine Fromageot, without whose unfailing enthusiasm and imaginative hard work the precursor to the present book would not have been undertaken. He thanks David Tranah of the Cambridge University Press who has made the publication of this book as easy and pleasant as possible, and Wagner Associates, of Skokie, Illinois, for their enthusiastic and perfect work in preparing the manuscript.

Roger Lyndon

Ann Arbor, 31 August 1984

CONTENTS

Chapter X. Fuchsian Groups

CHAPTER ONE: SYMMETRIES AND GROUPS

1. Definitions

A plane figure is any subset F of the Euclidean plane E.

A symmetry (isometry, rigid motion) of a figure F is a bijective (one-to-one) map α from F onto F that preserves distance: the distance $d(P,Q)$ between points P and Q of F is the same as the distance $d(P\alpha,Q\alpha)$ between their images $P\alpha$ and $Q\alpha$.

We write Sym F for the set of all symmetries of F.

2. Examples

I. A tree. Let F be a conventionalized picture of a tree, as shown. Then F has 'lateral symmetry' in the vertical line ℓ passing through the topmost point A. Explicitly F has a symmetry ρ that maps every point P of F that lies on ℓ to itself, and maps every other point P of F to a point $P\rho$ on the same horizontal as P, and at the same distance from ℓ, but on the other side of ℓ. The map ρ is a reflection with axis ℓ.

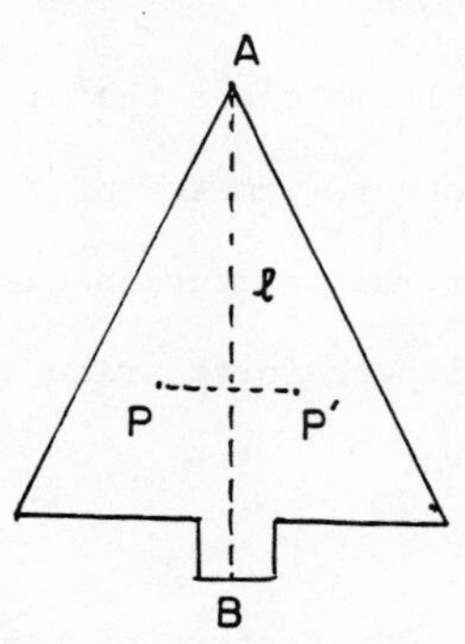

Every figure F has a trivial symmetry ε, mapping each point to itself. It is easy to see that ε and ρ are the only symmetries of F, whence Sym F = $\{\varepsilon,\rho\}$. One could prove this, for example, by arguing that F contains no other point like A, where the boundary of F makes the same angle as at A, and that F contains no other point like B. From this it follows that, if α is any symmetry of F, then $A\alpha = A$ and $B\alpha = B$. Now α must fix every point of F on the line ℓ, and, for each P not on ℓ, either $P\alpha = P$ or $P\alpha = P\rho$, with the same choice for all P. Thus either $\alpha = \varepsilon$ or $\alpha = \rho$.

If α and β are any two symmetries of a figure F, it is clear from the definition that α followed by β is also a symmetry of F. We write $\alpha\beta$ for this *product* of α followed by β; explicitly, for all points P of F, $P(\alpha\beta) = (P\alpha)\beta$. Thus a *multiplication* (*composition*) is defined on the set Sym F, and the set Sym F equipped with this multiplication becomes a *group*, the *symmetry group* Sym F of F. The *multiplication table* for the symmetry group Sym F of the tree is shown.

	ε	ρ
ε	ε	ρ
ρ	ρ	ε

It is clear that if ε is the trivial symmetry of a figure F and α is any other symmetry of F, then $\varepsilon\alpha = \alpha\varepsilon = \alpha$. The trivial symmetry ε acts under multiplication exactly as the number 1, and, for this reason, we shall henceforth write 1 rather than ε for the trivial symmetry of any figure.

II. Letters of the alphabet. We write the letters of the alphabet in a highly symmetric form.

ABCDEFGHIJKLM
NOPQRSTUVWXYZ

The letters A, M, T, U, V, W, and Y evidently admit exactly the same symmetries as the tree: the trivial (or <u>identity</u>) symmetry and reflection in a vertical axis.

The letters B, C, D, E, and K admit a single nontrivial symmetry π, reflection in a horizontal axis. The multiplication table, as shown, is the same as that of the tree, except that ρ has been replaced by π. The two groups are <u>isomorphic</u>: there is a bijective map Ω from one to the other that preserves <u>multiplication</u>: for any two symmetries α and β of the first, the product of their images is the image of their product. $(\alpha\Omega)(\beta\Omega) = (\alpha\beta)\Omega$.

	1	π
1	1	π
π	π	1

The letters F, G, J, L, P, Q, R have only the trivial symmetry. Their symmetry group is the trivial group, Sym F = {1}, or, in customary notation, Sym F = 1. The multiplication table is as shown.

	1
1	1

The letters N, S, and Z admit only one nontrivial symmetry σ, a <u>half turn</u>, or rotation through π, about the center of the figure. The symmetry group is again isomorphic to that of the tree. We emphasize that although these groups are isomorphic as 'abstract' groups, they are in an obvious sense geometrically different.

The remaining letters H, I, O, X, as drawn, admit all the symmetries 1, ρ, π, σ considered so far, and, in fact, admit no others. Thus, for these figures F, Sym F = $\{1, \rho, \pi, \sigma\}$, with multiplication table as shown. To show, for example, that these are all symmetries of the letter H, it suffices to examine what each symmetry α

	1	ρ	π	σ
1	1	ρ	π	σ
ρ	ρ	1	σ	π
π	π	σ	1	ρ
σ	σ	π	ρ	1

does to the four ends of the vertical lines in the letter H.

Note that if we regarded the letter I as an 'infinitely thin' vertical line segment without horizontal crosspieces, then the maps 1 and ρ would coincide as maps restricted to the points of the figure. Likewise π and σ would coincide, and the symmetry group would reduce to $\{1,\pi\}$. Usually it doesn't matter whether we regard the symmetries of a figure F as maps from F to F, or as maps from the entire plane E to itself which map the subset F to itself. The exceptions arise only in the case that the figure F is contained in some line in E.

Note that if we regarded the letter O as a circle, it would admit as symmetries all rotations, through any angle, about its center, as well as reflections in any diameter. The symmetry group would thus be infinite.

If we regarded the letter X as a cross, made up of two perpendicular line segments bisecting each other, then its symmetry group would contain eight elements.

III. An equilateral triangle. Let F be an equilateral triangle with vertices A, B, C. Let α, β, γ be reflections in the altitudes a, b, c through A, B, C; clearly α, β, γ are symmetries of F. Let σ be rotation through $2\pi/3$ about the center O of F; clearly σ, σ^2, and $\sigma^3 = 1$ are symmetries of F. In fact, these are all symmetries of F.

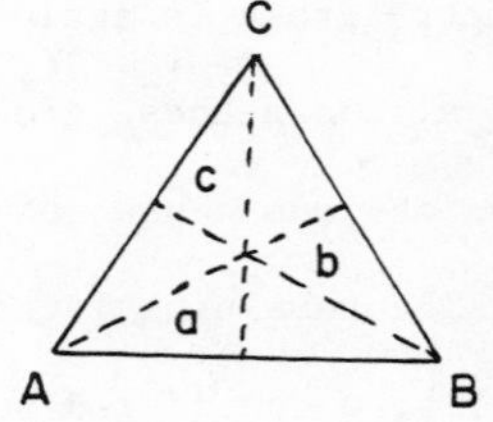

To prove this, observe that every symmetry of F must permute the set $V = \{A,B,C\}$ of vertices of F, and that, if two symmetries of F permute V in the same way they must be the same. Since there are exactly six permutations of a set V of three elements, Sym F cannot have more than the six elements that we have already found.

We shall not calculate the 6-by-6 multiplication table for Sym F; this is easy enough but rather tedious, and we shall find a simpler way to describe this group abstractly later. However, we illustrate the method by calculating the two products $\alpha\beta$ and $\beta\alpha$. Since α fixes A while interchanging (*transposing*) B and C, and β fixes B while transposing C and A, and γ fixes C while transposing A and B, we find that $A(\alpha\beta) = (A\alpha)\beta = A\beta = C$, that $B\alpha\beta = C\beta = A$, and that $C\alpha\beta = B\beta = B$. Thus $\alpha\beta$ permutes V in the same way as σ^2, and $\alpha\beta = \sigma^2$. A similar calculation gives $A\beta\alpha = C\alpha = B$, $B\beta\alpha = B\alpha = C$, and $C\beta\alpha = A\alpha = A$, whence $\beta\alpha = \sigma$. We emphasize that $\alpha\beta \neq \beta\alpha$. The group Sym F is *noncommutative* (*nonabelian*) and is in fact the smallest nonabelian group.

3. Groups

We now give precise definitions of some of the concepts used above.

Definition. A *group* is a pair of things, first a nonempty set G of objects called *elements*, and second an operation of *multiplication* associating with any two elements x and y of G a third element of G, written as xy and called their *product*. It is required that this multiplication satisfy three conditions:

(1) For all x, y, z in G, $(xy)z = x(yz)$;

(2) There is an element 1 in G such that, for all x in G,
$1x = x1 = x$;

(3) For all x in G there is an element x^{-1} in G such that
$x^{-1}x = xx^{-1} = 1$.

Remarks. (1) Although we have defined a group to be a set G together with an operation of multiplication defined on G, it is universal practice to speak simply of the group G, leaving the multiplication to be understood from the context.

(2) If G is a set of mappings and multiplication is defined by applying first one and then the other, then the associative law $(xy)z = x(yz)$ is automatically fulfilled.

(3) If G is a set of mappings and G contains the identity map 1, then the condition $1x = x1 = x$ is automatic.

(4) The existence of an inverse x^{-1} to a map x requires that x be bijective.

(5) It is easy to see that a group cannot contain more than one element e such that $ex = xe = x$ for all x, and that, for given x in a group, there cannot be more than one element y such that $yx = xy = 1$.

(6) Our axioms for a group are deliberately somewhat redundant.

Examples of groups. The set Sym F of all distance preserving maps from a set F to itself, for F a subset of any geometrical space, is a group. The set of all invertible linear transformations of any vector space is a group. The set of all permutations of any set Ω is a group, Sym Ω, the symmetric group on the set Ω. (No concept of distance enters here, but one can think of an abstract set as a 'space' in which all distances between distinct points are equal.)

Definition. A group H is a subgroup of a group G if H is a subset of G and the multiplication in H is the same as that in G when restricted to elements in H. More simply, a subset H of G is a subgroup if $1 \in H$, if $x \in H$ implies $x^{-1} \in H$, and if $x, y \in H$ implies $xy \in H$.

Definition. An isomorphism ϕ from a group G_1 to a group G_2 is a bijection from G_1 to G_2 that preserves multiplication: $(xy)\phi = (x\phi)(y\phi)$ for all $x, y \in G_1$. Two groups G_1 and G_2 are isomorphic if there exists an isomorphism from G_1 to G_2.

<u>Example</u>. We have seen that if F is an equilateral triangle and V is the set of its vertices, then Sym F and Sym V are isomorphic.

The following rather easy but important theorem clarifies the connection between the class of 'abstract' groups, as given by our definition, and the 'concrete' geometric groups which will interest us in these notes.

<u>CAYLEY'S THEOREM</u>. <u>Every group is isomorphic to a permutation group</u>.

<u>Proof</u>. Let a group G be given. We must choose a set V of objects to be permuted, and we choose this to be G itself, or, more precisely, the set of elements of G. Now the set $G = V$ will play two roles in the discussion, and, for this reason, we retain two names for it, G as the given group and V as the set of objects being permuted.

If g is any element of G and $v \in V$, then (since $V = G$), $vg \in V$. It is easy to check that the map (right multiplication by g) carrying each v to vg is a permutation of V; we call this permutation $g\phi$. We have thus defined a map ϕ from G to Sym V, carrying each $g \in G$ to $g\phi \in \text{Sym } V$.

If $g\phi = h\phi$, then, for $1 \in V$, $1(g\phi) = 1g = g$ must equal $1(h\phi) = 1h = h$, whence $g = h$. Thus ϕ is a bijection from G onto a subset $G\phi$ of Sym V. Any two elements of $G\phi$ are of the form $g\phi$ and $h\phi$ for some g, h in G; now, for any v in V, repeated application of the various definitions shows that

$$v\big((g\phi)(h\phi)\big) = \big(v(g\phi)\big)(h\phi) = (vg)(h\phi) = (vg)h = v(gh) = v\big((gh)\phi\big),$$

whence $(g\phi)(h\phi) = (gh)\phi$. We conclude, first, that the product of any two elements of $G\phi$ is again an element of $G\phi$, and deduce easily that $G\phi$ is a subgroup of Sym V, that is, a group of permutations.

We conclude, second, that the map ϕ preserves multiplication; therefore, as a bijection from G to $G\phi$, it is an isomorphism from G to the permutation group $G\phi$. □

Remarks. We have stated Cayley's Theorem in its simplest form. Clearly the proof contains considerable more detailed information.

Cayley's Theorem illustrates one of the central processes in mathematics. One begins with a variety of more or less concrete objects that arise in the daily practice of mathematics, here groups of permutations, or of transformations. Observing certain uniformities, one lists as 'axioms' certain properties that the objects have in common, and passes to the study of all 'abstract' objects satisfying these axioms. With good fortune, and good judgement, one is then able to prove that all objects satisfying these axioms are isomorphic to objects of the class of concrete objects (possibly enlarged by hindsight) in which one was originally interested.

The usefulness of this interplay between the concrete and the abstract is fairly obvious. In studying a variety of concrete objects, here various groups arising in geometry, we are able to unify our discussion within the abstract theory of all groups. In the other direction, it is equally important that, in developing the abstract theory of groups, we frequently can profit from representing an abstract group in concrete form.

4. Symmetries of regular polygons

A regular polygon is a polygon with $n \geqslant 3$ sides of equal length in which the interior angles, between adjacent sides, are all equal. For $n = 3$, the regular polygon is an equilateral triangle; for $n = 4$ it is a square. We have described already the symmetry group of the equilateral triangle, and the symmetry group Sym F of the regular polygon F of $n > 3$ sides follows the same pattern.

Let F be a regular polygon with $n \geqslant 3$ sides, and let G = Sym F. It is clear that G contains a rotation σ about the center of F through an angle of $2\pi/n$; moreover, $\sigma^1 = \sigma, \sigma^2, \ldots, \sigma^{n-1}$ are all distinct, while $\sigma^n = \sigma^0 = 1$. Thus G contains n rotational symmetries $1, \sigma, \sigma^2, \ldots, \sigma^{n-1}$.

By looking at what happens to the n vertices of F, it is clear that these are all the symmetries of F that <u>preserve orientation</u>, that is, preserve the cyclic order of vertices around the polygon, or do not 'turn the polygon over'.

It turns out that all other symmetries of F are reflections. At each vertex P there is an 'altitude' or 'diameter' p of F, passing through the center 0 of F; if n is odd, p bisects a side of F opposite P, while if n is even p ends at a vertex Q opposite P. If n is even, there is also a line through 0 bisecting two opposite sides. Whether n is odd or even, there are in all exactly n such 'diameters', which are clearly axes for reflectional symmetries of F.

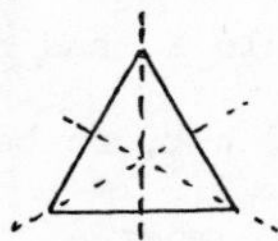
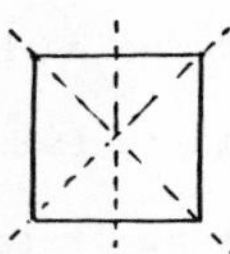
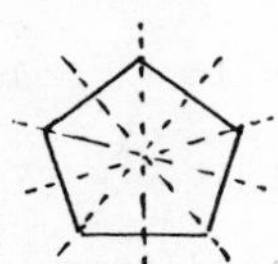

We prove that G consists of exactly these 2n elements. First, if α is an orientation preserving symmetry of F, then $\alpha = \sigma^k$ for some integer k. Let ρ be any one of the n reflectional symmetries described above. If α reverses orientation, then, since ρ also reverses orientation, $\rho\alpha$ preserves orientation. It follows that $\rho\alpha = \sigma^k$ for some k, and that $\alpha = \rho^{-1}\sigma^k$ or, since $\rho^2 = 1$, that $\alpha = \rho\sigma^k$. We have proved the following:

(1) G <u>consists of exactly</u> 2n <u>elements</u>: n <u>rotations</u> $1, \sigma, \sigma^2, \ldots, \sigma^{n-1}$ <u>and</u> n <u>reflections</u> $\rho, \rho\sigma, \rho\sigma^2, \ldots, \rho\sigma^{n-1}$.

(2) G <u>is generated by</u> σ <u>and</u> ρ: <u>every element of</u> G <u>can be written as a product of powers of</u> σ <u>and powers of</u> ρ.

Next, rather than describing a multiplication table for G, we describe <u>rules</u> for multiplying elements of G. First, from the equations $\sigma^n = 1$ and $\rho^2 = 1$, we are clearly permitted to reduce the exponent k on an element σ^k modulo n, and the exponent h on an element ρ^h modulo 2. In addition to the two <u>relations</u> $\sigma^n = 1$ and $\rho^2 = 1$, each involving only one of the <u>generators</u> σ and ρ, we verify a further relation $\rho\sigma\rho = \sigma^{-1}$ involving both of them, by direct calculation, by intuition, or from experience, say driving a screw into the underside of a table. From this relation it follows that $\sigma\rho = \rho\sigma^{-1}$ and, generally, $\sigma^k\rho = \rho\sigma^{-k}$.

We now show that the three relations above form a set of <u>defining relations</u> in the sense that if w_1 and w_2 are any two <u>words</u> in the generators σ and ρ, that is, products of powers of σ and ρ, and if w_1 and w_2 represent the same element of the group G, then the equation $w_1 = w_2$ follows from the three relations. More explicitly, we show that w_1 and w_2 can be reduced to the same form by repeated application of the rules of reducing k in σ^k modulo n, of reducing h in ρ^h modulo 2, and of replacing any part $\sigma^k\rho$ by $\rho\sigma^{-k}$. This is fairly obvious, but, to be precise, we reason by induction, at each stage diminishing the number of parts of the form $\sigma^k\rho^h$. If no such part occurs, w_1 (or w_2) is in one of the 2n canonical forms σ^k of $\rho\sigma^k$ for $k = 0,1,\ldots,n-1$. If such a part does occur, we may suppose first of all that all parts ρ^h are of the form ρ. Now we must have either $w_i = \sigma^k\rho$ or w_i contains $\rho\sigma^k\rho$; applying the rule $\sigma^k\rho = \rho\sigma^{-k}$ either gives $w_i = \rho\sigma^{-k}$ or replaces the part $\rho\sigma^k\rho$ by $\rho^2\sigma^{-k} = \sigma^{-k}$. This completes the induction, and shows that both w_1 and w_2 are reducible to canonical form. Now, if w_1 and w_2 represent the same element of G, these two canonical forms must be the same, and we have established the equation $w_1 = w_2$.

We summarize this result by saying that G has a presentation, by generators and relations, of the form

$$G = \langle \sigma, \rho : \sigma^n = 1,\ \rho^2 = 1,\ \rho\sigma\rho = \sigma^{-1} \rangle .$$

This is clearly more concise than giving a multiplication table, and, in most cases it gives more easily usable information. Moreover, if the group is infinite, it is not possible to give a multiplication table. In practice, we shall describe groups in two ways: either a geometric description, or a presentation by generators and relations.

A group will have infinitely many different presentations, and one ordinarily tries to choose one that is reasonably simple and which exhibits clearly various properties of the group. For example, in the presentation above, the third relation can be rewritten as $(\rho\sigma)^2 = 1$. The order of an element x in a group G is defined to be the least positive integer n such that $x^n = 1$, if one exists, and otherwise is said to have infinite order. Now the presentation

$$G = \langle \sigma, \rho : \sigma^n = 1,\ \rho^2 = 1,\ (\rho\sigma)^2 = 1 \rangle$$

can be described in words as saying that G is (the 'most general' group) generated by elements of orders n and 2 whose product has order 2.

If we write $\rho_1 = \rho$ and $\rho_2 = \rho\sigma$, then $\sigma = \rho_1\rho_2$, and the presentation can be rewritten in the form

$$G = \langle \rho_1, \rho_2 : \rho_1^2 = 1,\ \rho_2^2 = 1,\ (\rho_1\rho_2)^n = 1 \rangle .$$

This displays G as generated by two reflections whose product has order n.

It is clear that the set G^+ of elements of G that preserve orientation is a subgroup of G, and G^+ evidently has a presentation

$$G^+ = < \sigma : \sigma^n = 1 > .$$

This group is called, for obvious reasons, a <u>cyclic group</u> C_n, of <u>order</u> (number of elements) n. By analogy, the group generated by a single element of infinite order, $C_\infty = < \sigma : \emptyset >$, (with no relations), is the <u>infinite cyclic group</u>.

The group G = Sym F, for F a regular polygon of $n \geqslant 3$ sides is the <u>dihedral group</u> of order 2n; we shall denote it as D_{2n} although, unfortunately, the notation D_n is equally common. We extend the notation to the cases $n = 1, 2, \infty$ in the obvious way:

$$D_2 = < \rho : \rho^2 = 1 > ;$$
$$D_4 = < \sigma,\rho : \sigma^2 = 1,\ \rho^2 = 1,\ (\rho\sigma)^2 = 1 > ;$$
$$D_\infty = < \sigma,\rho : \rho^2 = 1,\ (\rho\sigma)^2 = 1 > ,$$

or, alternatively,

$$D_\infty = < \rho_1, \rho_2 : \rho_1^2 = 1,\ \rho_2^2 = 1 > .$$

We also define $C_n = < \sigma : \sigma^n = 1 >$ for all $n \geqslant 1$, with $C_\infty = < \sigma : \emptyset >$.

Evidently D_2 is isomorphic to C_2. Moreover, the group D_4 has a presentation

$$D_4 = < \sigma,\rho : \sigma^2 = 1,\ \rho^2 = 1,\ \sigma\rho = \rho\sigma > ,$$

whence it is abelian. This abelian group of order 4 is often referred to as (Felix Klein's) <u>four-group</u> (Vierergruppe). We have encountered it before as the symmetry group of the letter H.

5. Presentations

The concept of a presentation of a group is fairly simple. A group G is generated by a set X of elements, and all relations $w_1 = w_2$ between products w_1, w_2 of powers of elements of X are consequences of the given set of defining relations. If the reader is content with this definition he may omit the following discussion. Nonetheless, it may be instructive to sketch how this concept can be formalized within group theory, and without appeal to the somewhat extraneous idea of logical 'consequence'.

We begin by constructing a free group F with a given set X as basis. This is to be the group $F = \langle X : \emptyset \rangle$ with the set X of generators and with no defining relations. Each element of F will be expressible as a word $w = x_1 \ldots x_n$, $n \geqslant 0$, where each $x_i = x^{\pm 1}$ for some x in X. If, similarly, $u = y_1 \ldots y_m$, $m \geqslant 0$, each $y_i \in X^{\pm 1}$, the product is $wu = x_1 \ldots x_n y_1 \ldots y_m$. Clearly two words w and w' represent the same element of F if one can be obtained from the other by repeated insertion or deletion of parts xx^{-1}, for x in $X^{\pm 1}$; in this case we call w and w' equivalent and write $w \equiv w'$. Let F be the set of equivalence classes [w] of words. If $w_1 \equiv w_1'$ and $w_2 \equiv w_2'$, then $w_1 w_2 \equiv w_1' w_2'$, whence we can define unambiguously a multiplication in F by setting $[w_1] \cdot [w_2] = [w_1 w_2]$. It is easy to see that, with this multiplication, F becomes a group, the free group F with basis X.

The elements of F, as defined, are equivalence classes [w] of words, and the words w may be viewed as names for elements of F. It is customary to pass over this distinction, speaking of 'the element w of F'.

The distinctive property of the free group F with basis X is contained in the following theorem.

THEOREM. *Let X be any set, G any group, and χ any map from the set X into the group G. Then there exists a unique homomorphism ϕ from the free group F with basis X into G with the property that $x\phi = x\chi$ for all x in X.*

Proof. If any such map ϕ exists, we must have $(x^{\pm 1})\phi = (x\chi)^{\pm 1}$ for all x in X, and, for $w = x_1 \ldots x_n$, we must have $w\phi = (x_1\phi)\ldots(x_n\phi)$. One verifies that $w \equiv w'$ implies $w\phi = w'\phi$. This determines a map ϕ from F to G uniquely, and it is easy to see that $(ww')\phi = (w\phi)(w'\phi)$, whence ϕ is indeed a homomorphism. □

COROLLARY. _Let a group_ G _be generated by some subset_ Y _of its elements, and let_ X _be any set in bijective correspondence_ $\chi : X \to Y$ _with_ Y. _Then there exists a homomorphism_ ϕ _from the free group_ F _with basis_ X _onto_ G, _such that_ $x\phi = x\chi$ _for all_ x _in_ X.

Proof. It is necessary only to notice that since the image $F\phi$ contains the set $X\phi = Y$, which generates G, $F\phi$ must be all of G. □

Quite generally, if ϕ is any homomorphism from a group F onto a second group G, the _kernel_ N of ϕ, consisting of all u in F such that $u\phi = 1$, is a subgroup of F and, indeed, a _normal subgroup_: $u \in F$, $r \in N$ implies that $u^{-1}ru \in N$. If N is any normal subgroup of a group F, then the set F/N of all _cosets_ $Nu = \{ru: r \in N\}$ form a group: $(Nu)(Nv) = \{u'v': u' \in Nu, v' \in Nv\} = N(uv)$. If N is the kernel of ϕ from F onto G, then setting $(Nu)\bar{\phi} = u\phi$ defines an isomorphism $\bar{\phi}$ from F/N onto G.

We return to a homomorphism ϕ from a free group F with basis X onto a group G. Now two elements w_1 and w_2 of F represent the same element of G if and only if $w_1\phi = w_2\phi$; in this case we say that the _relation_ $w_1 = w_2$ _holds in_ G. Since $w_1 = w_2$ is equivalent to $w_2^{-1}w_1 = 1$, it suffices to consider relations of the form $w = 1$. Now a relation $w = 1$ holds in G if and only if $w\phi = 1$, that is, if and only if $w \in N$, the kernel of ϕ; the elements $w \in N$ are called _relators_.

A set of relations $r = 1$ is called a _set of defining relations_ (or, better, a defining set of relations) if all other relations are consequences of them. A subset R of F is called a _set of defining relators_ if the relations $r = 1$, for $r \in R$, form a set of defining relations. This is equivalent to the condition that N be the normal closure of R in F, that is, the least normal subgroup of F containing R, or, more explicitly, N is the set of all products of powers of conjugates $u^{-1}ru$ of elements $r \in R$ by elements $u \in F$.

We are at last in a position to give a formal definition of a presentation of a group G.

DEFINITION. *A presentation of a group* G *is a triple* (ϕ, X, R) *where* ϕ *is a homomorphism from the free group* F *with basis* X *onto* G *and the kernel* N *of* ϕ *is the normal closure of the subset* R *of* F.

Having made this definition, we shall not mention it again. We revert to our more informal language; however, whenever we say that some property of presentations is 'obvious' or 'can easily be shown', we mean implicitly that it can be derived easily from this formal definition.

However, we shall retain the notation $G = \langle X: R \rangle$ as an alternate to the previous notation $G = \langle X: \{r = 1: \text{all } r \in R\} \rangle$.

6. Change of presentation

For simplicity we confine our discussion to finite presentations $G = \langle X : R \rangle$, that is, such that both X and R are finite.

There are four rather obvious ways of changing one presentation of a group to another, and we shall see that any two finite presentations of the same group are related by a finite succession of such changes. These changes are called Tietze transformations and are of four types, which we now describe.

Type 1 consists in replacing a presentation $\langle X : R \rangle$ by a presentation $\langle X: R \cup \{r\} \rangle$ where r is a consequence of R, that is, r is in the normal closure of R. (This amounts to adding a redundant defining relation.) Type 1* is the opposite transformation, passing from $< X: R \cup \{r\} >$ to $< X : R >$, where r is a consequence of R (deleting a redundant defining relation).

Type 2 consists in replacing $< X : R >$ by $< X \cup \{x\}:R \cup \{r\} >$ where x is a new element not in X and r has the form $x^{-1}w$ for w some element of the free group F with basis X. (This amounts to adding a new generator x together with a defining relation $x = w(x_1,\ldots,x_n)$ that defines x in terms of the generators $x_1,\ldots,x_n$ already present in X.) Type 2* is the opposite transformation, from $< X \cup \{x\}:R \cup \{r\} >$ to $< X : R >$ under the same circumstances, and amounts to eliminating a redundant generator together with the equation expressing it in terms of the remaining generators.

It should be intuitively obvious, and is easily proved, that these transformations do not change the group G presented.

THEOREM. Given any two finite presentations of the same group one can pass from one to the other by a finite succession of Tietze transformations.

Proof. Let $< X : R >$ and $< Y : S >$ be two finite presentations of the same group G.

Since (the image of) X generates G, for each $y \in Y$ there must hold some relation $y = w_y(x_1,\ldots,x_n)$, a word in the free group with basis X. For each $y \in Y$ we choose such a w_y and define U to be the set of all $u_y = y^{-1}w_y$. By a succession of transformations of type 1 we can pass from $< X : R >$ to the presentation $< X \cup Y: R \cup U >$.

Similarly, for each $x \in X$ some relation $x = w_x$ holds in G for $w_x = w_x(y_1,\ldots,y_m)$ a word in the free group with basis Y, and we let V be the set of all $v_x = x^{-1}w_x$.

Now all the relations $v_x = 1$ hold in G, hence must be consequences of the set $R \cup U$ of defining relations. By a succession of transformations of type 1 we can pass from $< X \cup Y: R \cup U >$ to $< X \cup Y: R \cup U \cup V >$. Similarly, for all $s \in S$, the relation $s = 1$ holds in G, hence is a consequence of the set $R \cup U \cup V$, and, by further transformations of type 1 we can pass to $< X \cup Y: R \cup S \cup U \cup V >$.

We have shown that, by a succession of Tietze transformations, we can pass from $< X : R >$ to $< X \cup Y: R \cup S \cup U \cup V >$. The same argument shows that we can pass from $< Y : S >$ to $< X \cup Y: R \cup S \cup U \cup V >$. Since Tietze transformations are, by definition, reversible, we can pass from $< X \cup Y: R \cup S \cup U \cup V >$ to $< Y : S >$. We have shown that one can pass from $< X : R >$ to $< Y : S >$ via $< X \cup Y: R \cup S \cup U \cup V >$. □

Example. We show that the cyclic group C_6 is the direct product $C_2 \times C_3$ of cyclic groups C_2 of order 2 and C_3 of order 3, that is, that the two presentations $< a: a^6 = 1 >$ and $< b,c: b^2 = 1, c^3 = 1, bc = cb >$ define isomorphic groups. Of course, this could be done directly, but here we give a proof in terms of Tietze transformations.

To conform to the usage above, we rewrite the presentations in the forms $< a: a^6 >$ and $< b,c: b^2, c^3, bcb^{-1}c^{-1} >$. We start with the presentation $< a: a^6 >$ and define $b = a^3$, $c = a^2$, that is, by two steps of type 2 we pass to the presentation $< a,b,c: a^6, b^{-1}a^3, c^{-1}a^2 >$. It is clear that these relations imply that $b^2 = 1$, $c^3 = 1$, $bc = cb$, and also $a = bc^{-1}$, whence we may add to our presentation the redundant relators b^2, a^3, $bcb^{-1}c^{-1}$, $a^{-1}bc^{-1}$. (One can check (unnecessarily) that, for example, $a^{-1}bc^{-1} = (a^2(b^{-1}a^3)a^{-2})^{-1}(a^2(c^{-1}a^2)a^{-2})$, in the normal closure of $b^{-1}a^3$ and $c^{-1}a^2$.) Thus we have a presentation $< a,b,c: a^6, b^{-1}a^3,$

$c^{-1}a^2, b^2, c^3, bcb^{-1}c^{-1}, a^{-1}bc^{-1}$ >. It is again clear that the relations $b^2 = 1$, $c^3 = 1$, $bc = cb$, and $a = bc^{-1}$ imply the relations $a^6 = 1$, $b = a^3$, and $c = a^2$. (This again could be checked formally.) Thus we can delete the first three relators to obtain a presentation $\langle a,b,c: b^2, c^3, bcb^{-1}c^{-1}, a^{-1}bc^{-1} \rangle$. Finally, by a step of type 2* we can delete the generator a together with the relator $a^{-1}bc$, to obtain $\langle b,c: b^2, c^3, bcb^{-1}c^{-1} \rangle$, as desired.

REMARK. One should be warned that this theorem does not provide an algorithm for deciding whether two given finite presentations define isomorphic groups. This follows from the fact that the <u>word problem for groups</u> is unsolvable: there exists no algorithm which, given any free group F with a finite basis X, a finite subset R of F, and an element r of F, decides whether r belongs to the normal closure of R in F. In fact, the <u>triviality problem</u> is unsolvable: there is no algorithm which, given any finite presentation, decides whether the group it defines is a trivial, one-element, group. However, these unsolvability results do not seriously diminish the usefulness of Tietze transformations in attempting to establish the equivalence of two presentations when one has good reason to believe that they define isomorphic groups. (We hope also that they offer some justification for giving a rigorous formal definition of a presentation.)

Notes and Problems

Problem 1. What is the symmetry group of a rectangle with unequal dimensions (not a square)? What is the symmetry group of a rectangular solid (brick) with three unequal dimensions a, b, c? What if $a = b \neq c$?

Problem 2. A regular tetrahedron is a pyramid with triangular base, all of whose faces (base and three sloping faces) are congruent equilateral triangles. What is its symmetry group G? Classify the elements of G according to their sets of fixed points. (Hint: if α fixes P and carries a vertex V_1 to a vertex V_2, then P is equidistant from V_1 and V_2.) Which elements are rotations? Which elements are reflections in planes?

Problem 3. Find all subgroups of D_{2n} and of D_∞.

Problem 4. Let F be the infinite set of all points (n,0) in the coordinate plane such that n is an integer. What is Sym F?

Problem 5. Show that the symmetry group of a circle is not generated by any finite number of elements. Obtain a description of it by generators and relations, analogous to that for D_{2n}.

Problem 6. For p a prime number, show that all groups of order p (with p elements) are isomorphic. Show that there are exactly two isomorphism classes of groups of order 4, and 2 of groups of order 6.

Note 1. The free product $G = G_1 \star G_2$ of two groups G_1 and G_2 can be defined as follows. Let $G_1 = \langle X_1 : R_1 \rangle$ and $G_2 = \langle X_2 : R_2 \rangle$ (where it is assumed that X_1 and X_2 are disjoint); then $G = \langle X_1 \cup X_2 : R_1 \cup R_2 \rangle$. For example, the infinite dihedral group $D_\infty = \langle \rho_1, \rho_2 : \rho_1^2, \rho_2^2 \rangle = G_1 \star G_2$ for $G_1 = \langle \rho_1 : \rho_1^2 \rangle$ and $G_2 = \langle \rho_2 : \rho_2^2 \rangle$, the free product of two groups of order 2. The modular group, a very important group that we will encounter later, is the free product of a group of order 2 with a group of order 3.

Problem 7. Show that the elements of D_∞ are exactly the products $w = w_1 \ldots w_n$, $n \geqslant 0$, where each w_i is ρ_1 or ρ_2 and no $w_i = w_{i+1}$. Show that if n is even and positive then w has infinite order, while if n is odd w has order 2. For the modular group $G = \langle a, b : a^2, b^3 \rangle$ show that the elements are exactly the products $w = w_1 \ldots w_n$, $n \geqslant 0$, where each w_i is a, b, or b^{-1} and no $w_i = w_{i+1}^{\pm 1}$. Conclude that every element has order 1, 2, 3 or infinite order.

Note 2. The direct product $G = G_1 \times G_2$ of two groups G_1 and G_2 can be obtained from the free product by adding relations $g_1 g_2 = g_2 g_1$ for all g_1 in G_1 and g_2 in G_2. More simply, it can be defined as the set of all ordered pairs (g_1, g_2) for g_1 in G_1 and g_2 in G_2, with multiplication $(g_1, g_2)(g_1', g_2') = (g_1 g_1', g_2 g_2')$. We have seen that $C_6 \simeq C_2 \times C_3$.

Problem 8. Show that if n = ab where a and b are relatively prime positive integers, then $C_n \simeq C_a \times C_b$. Show that C_4 is not isomorphic to $C_2 \times C_2$. Show that if G_1 and G_2 are two normal subgroups of a group G, with trivial intersection $G_1 \cap G_2 = 1$, that together generate G, then $G \simeq G_1 \times G_2$.

Problem 9. The presentation $G = \langle \sigma, \rho : \sigma^n = \rho^2 = (\rho\sigma)^2 = 1 \rangle$ requires that σ, ρ, and ρσ have orders dividing n, 2, and 2, but does not automatically ensure that they have exactly these orders. This could be proved geometrically, since we have proved that G is the dihedral group, by looking at the action of σ, ρ, and ρσ on the set V of vertices of the n-gon. Alternatively, we could use the method used to prove Cayley's theorem to represent G as a permutation group. For this, let V be a set of 2n elements u_k and v_k, for $k = 0,\ldots,n-1$ (we have in mind the elements σ^k and $\rho\sigma^k$). Define permutations $\sigma,\rho \in \text{Sym } V$ by specifying $u_k\sigma$, $v_k\sigma$, $u_k\rho$, $v_k\rho$ in the obvious way (for example, $u_k\sigma = u_{k'}$ for $k' \equiv k + 1$ modulo n). Now verify that σ and ρ satisfy the given relations and that σ, ρ, $\rho\sigma$ have orders exactly n, 2, 2.

Problem 10. Consider groups G generated by elements σ and ρ satisfying the relations $\sigma^n = \rho^2 = (\rho\sigma)^2 = 1$. Show that these groups are exactly the trivial group G = 1 together with (groups isomorphic to) the dihedral groups D_{2m} for all $m \geqslant 1$ that divide n. Describe the isomorphism types of all groups G generated by elements x and y such that $x^2 = y^2 = 1$.

Note 3. The method used to show that the given relations form a set of defining relations for the dihedral group $G = D_{2n}$ provides a solution of the word problem for this group. It gives an algorithm for reducing each word w to canonical form σ^k or $\rho\sigma^k$. Given two words, we have only to check whether they have the same canonical form. Alternatively, we could appeal to the geometry, and check whether two given words permute the vertices of the n-gon in the same way.

However, it should be noted that there exist finite presentations for which the word problem is undecidable: there exists no algorithm for deciding whether an arbitrary pair of words represent the same element of G.

References

For a broad nontechnical discussion of symmetry we recommend the book Symmetry by H. Weyl (see bibliography at the end of these notes). More specialized discussions of symmetry in geometry will be cited later.

For the reader not at ease with the material on groups in Section 3, we recommend reading the first one or two chapters of almost any introductory book on group theory, perhaps followed by judicious browsing. For a very readable introduction to the standard ideas of group theory, we especially recommend the first part of the book of J. Rotman. The book of D. L. Johnson is more specialized, dealing in particular with presentations, and more in the spirit of these notes, but is careful and readable, with a great abundance of examples.

Anyone wanting to know more about abstract infinite groups than is needed here should look at the careful and thorough discussion in the book of W. Magnus, A. Karrass, and D. Solitar. An accessible, but necessarily not always easy, discussion of the word problem and related matters is given in the last chapters of Rotman's book.

CHAPTER TWO: ISOMETRIES OF THE EUCLIDEAN PLANE

1. Geometric types of isometries

Let $\mathcal{E}$ be the group of all isometries of the Euclidean plane E, and let $\mathcal{E}^+$ be the subgroup of those isometries that preserve orientation. We begin with a geometrical classification of isometries of the plane.

A translation τ is a transformation that moves every point the same distance in the same direction. It is convenient to count the trivial map 1 as a translation. If τ is a nontrivial translation, and P and Q are any two points, then the directed segments (or vectors) $\overrightarrow{P,P\tau}$ and $\overrightarrow{Q,Q\tau}$ are parallel and of equal length $|\tau|$. Evidently a nontrivial translation fixes no point, and leaves invariant only those lines that contain some point P together with its image $P\tau$.

A rotation σ is an orientation preserving map that fixes some point O. If σ is not trivial, then O is unique, the center of σ. If $\sigma \neq 1$, then the angle θ from $\overrightarrow{O,P}$ to $\overrightarrow{O,P\sigma}$ is the same for all $P \neq O$, the turning angle of σ. A nontrivial rotation fixes no point other than O, and leaves no line invariant except in the case that $\sigma^2 = 1$, when σ maps each line through O onto itself with reversed direction.

A reflection ρ is a nontrivial isometry that fixes all points of some line ℓ, the axis of ρ. A point P not on ℓ is mapped to the point $P\rho$ such that ℓ perpendicularly bisects the segment $[P,P\rho]$. Evidently ρ reverses orientation, and $\rho^2 = 1$. The fixed points of ρ are exactly the points of ℓ, and the lines invariant under ρ are ℓ together with all lines ℓ' perpendicular to ℓ, which are reversed by ρ.

The remaining type of transformation is less familiar. A glide reflection ρ is a transformation $\rho = \rho_0\tau_0$ consisting of a reflection ρ_0 in an axis ℓ followed by a nontrivial translation τ_0 parallel to ℓ (that is, leaving ℓ invariant). Evidently ρ reverses orientation. Since ρ_0 and τ_0 clearly commute, $\rho = \rho_0\tau_0 = \tau_0\rho_0$ and $\rho^2 = (\tau_0\rho_0)(\rho_0\tau_0) = \tau_0\rho_0^2\tau_0 = \tau_0^2$, a nontrivial translation. Clearly ρ fixes no point and leaves invariant no line except its axis ℓ.

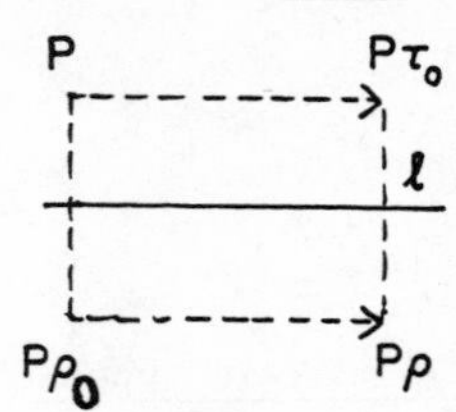

THEOREM. An isometry of the plane that fixes a point is a rotation if it preserves orientation and a reflection if it reverses orientation. An isometry of the plane that fixes no point is a translation if it preserves orientation and a glide reflection if it reverses orientation.

Proof. Suppose that α fixes a point 0. Then α maps any circle with center 0 onto itself, and hence is either a rotation or a reflection.

Suppose that α fixes no point. We note that α^2 also fixes no point, since if $P = P\alpha^2$, then α would fix the midpoint of the segment $[P,P\alpha]$.

Let α preserve orientation. If the vectors $\overrightarrow{P,P\alpha}$ and $\overrightarrow{P\alpha,P\alpha^2}$ are not parallel, their perpendicular bisectors meet in a point Q, and the triangles $[P,P\alpha,Q]$ and $[P\alpha,P\alpha^2,Q]$ are similar. Then α, carrying the side $[P,P\alpha]$ to the side $[P\alpha,P\alpha^2]$, and preserving orientation, would map Q to itself. We conclude that $\overrightarrow{P,P\alpha}$ and $\overrightarrow{P\alpha,P\alpha^2}$ are parallel, whence the distinct points P, $P\alpha$, $P\alpha^2$ lie on a line ℓ. It follows that all $P\alpha^k$ lie on ℓ, and that α acts on ℓ like a translation with vector $\overrightarrow{P,P\alpha}$, and, since it preserves orientation, that α acts

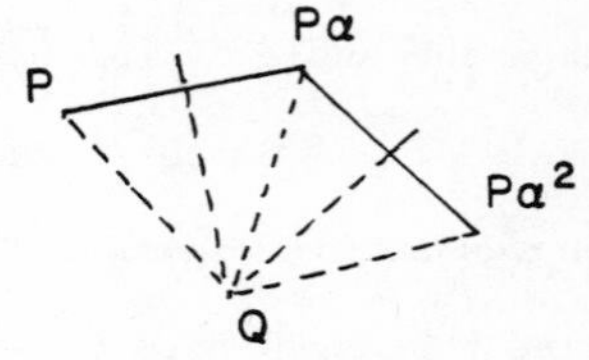

as this translation on all of the plane.

If α does not preserve orientation, then α^2 does, and the above argument applied to α^2 shows that α^2 is a nontrivial translation. Now α^2 leaves invariant the two lines $\ell_0 = \overline{P,P\alpha^2}$ and $\ell_1 = \overline{P\alpha,P\alpha^2}$, which are therefore parallel (but not necessarily distinct). Moreover, α interchanges ℓ_0 and ℓ_1, whence α leaves invariant the line ℓ parallel to ℓ_0 and ℓ_1 and half way between them. Since α^2 translates ℓ, α also must act as a translation τ_0 on ℓ. Finally, since α does not preserve orientation, it must be the glide reflection $\alpha = \rho_0\tau_0$, where ρ_0 is reflection in ℓ. □

THEOREM. *The product of two reflections with parallel axes is a translation perpendicular to these axes by a distance twice that from the first axis to the second. The product of two reflections with axes meeting in a point is a rotation about that point through an angle twice that from the first axis to the second.*

Proof. If reflections ρ_1,ρ_2 have parallel axes ℓ_1, ℓ_2, then the orientation preserving map $\tau = \rho_1\rho_2$ leaves invariant all lines ℓ perpendicular to ℓ_1 and ℓ_2, hence must be a translation. In the figure, $d(P_1,P_1\tau) = 2d(P_1,P_2)$.

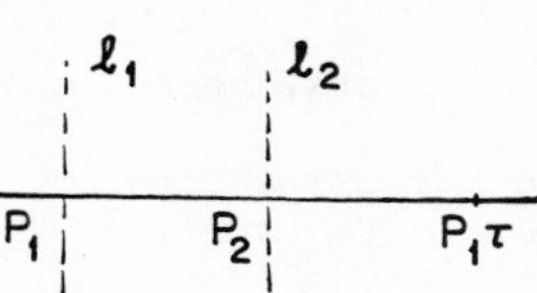

If ℓ_1 and ℓ_2 meet in 0, then the orientation preserving transformation $\sigma = \rho_1\rho_2$ fixes 0, hence must be a rotation about 0. In the figure, the angle $P_10(P_1\sigma)$ is twice the angle P_10P_2. □

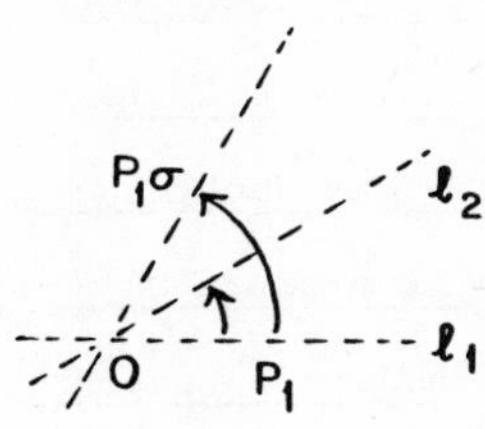

COROLLARY. <u>Every orientation preserving transformation is a product of two reflections, and every orientation reversing transformation is a product of three reflections. Thus, E is generated by reflections.</u>

<u>Proof</u>. We have seen that a rotation is a product of two reflections with axes through its center, and a translation is the product of two reflections with axes perpendicular to the direction of translation. A reflection ρ satisfies $\rho^2 = 1$ whence $\rho = \rho\rho\rho$. A glide reflection $\rho = \rho_0\tau_0$ is a product $\rho = \rho_0\rho_1\rho_2$ where ρ_0 is a reflection and $\tau_0 = \rho_1\rho_2$, a translation, is the product of two reflections ρ_1 and ρ_2. □

THEOREM. <u>The product of three reflections is a reflection if the three axes are parallel or concurrent, and otherwise is a glide reflection</u>.

<u>Proof</u>. If the three axes are parallel, then the product leaves invariant all lines perpendicular to the axes and, since it reverses orientation, it must be a reflection. If the axes meet in a point, the product must fix this point, and hence be a reflection. Conversely, suppose that the orientation reversing product $\rho = \rho_1\rho_2\rho_3$ is a reflection. Then $\rho_1\rho_2 = \rho\rho_3$. If $\rho_1\rho_2 = \rho\rho_3$ is a translation τ, then the axes of all of ρ_1, ρ_2, ρ_3, ρ must be perpendicular to τ, and hence parallel. If $\rho_1\rho_2 = \rho\rho_3$ is a rotation with center 0, then all the axes must pass through 0. □

THEOREM. (1) <u>The product of two translations is a translation</u>.

(2) <u>The product of two rotations is a rotation except if the sum of their angles is 0 modulo 2π, in which case it is a translation</u>.

(3) <u>The product of a translation and a nontrivial rotation is a rotation</u>.

(4) <u>The product of a nontrivial translation and a reflection is a glide reflection except if the axis of the reflection is perpendicular to</u>

<u>the translation, in which case it is a reflection.</u>

(5) <u>The product of a nontrivial rotation and a reflection is a glide reflection except when the axis of the reflection passes through the center of the rotation, in which case it is a reflection.</u>

<u>Proof</u>. (1) The product $\tau_1\tau_2$ of two translations τ_1 and τ_2 preserves orientation, and hence is a translation unless it fixes some point P. But $P\tau_1\tau_2 = P$ implies that $P\tau_1 = P\tau_2^{-1}$, whence $\tau_1 = \tau_2^{-1}$ and $\tau_1\tau_2 = 1$, the trivial translation.

(2) The product $\sigma_1\sigma_2$ of two rotations σ_1 and σ_2 with (positively measured) angles θ_1 and θ_2 carries any directed line $\vec{\ell}$ into a directed line $\vec{\ell}'$ making an angle $\theta_1 + \theta_2$ with $\vec{\ell}$. Since $\sigma_1\sigma_2$ preserves orientation, it is a translation if $\theta_1 + \theta_2 \equiv 0$ (modulo 2π), and is a rotation otherwise.

(3) The product $\tau_1\sigma_2$ preserves orientation and carries any $\vec{\ell}$ to $\vec{\ell}'$ making an angle of $\theta_2 \neq 0$ with ℓ.

(4) Let $\gamma = \tau\rho_3$, the product of a nontrivial translation τ and a reflection ρ_3 with axis ℓ_3. Write $\tau = \rho_1\rho_2$, the product of reflections ρ_1 and ρ_2 with axes ℓ_1 and ℓ_2 perpendicular to τ, and apply the preceding theorem.

(5) Analogous to (4), writing $\sigma = \rho_1\rho_2$ with axes ℓ_1, ℓ_2 through the center 0 of σ. □

THEOREM. <u>Let</u> α <u>be any element of</u> E.

(1) <u>If</u> τ <u>is a translation with an invariant line</u> ℓ, <u>then</u> $\tau^\alpha = \alpha^{-1}\tau\alpha$ <u>is a translation of equal length</u> $|\tau^\alpha| = |\tau|$ <u>with invariant line</u> $\ell\alpha$.

(2) <u>If</u> σ <u>is a rotation with center</u> 0, <u>then</u> σ^α <u>is a rotation with center</u> 0α <u>through the same angle as</u> σ.

(3) <u>If</u> ρ <u>is a reflection with axis</u> ℓ, <u>the</u> ρ^α <u>is a reflection with axis</u> $\ell\alpha$.

Proof. These assertions all follow directly from the observation that, if γ maps P to $P\gamma = Q$, then γ^{α} maps $P\alpha$ to $Q\alpha$. □

COROLLARY. The set of all translations is a subgroup T of E, and is normal in E, that is, $\tau \in T$ and $\alpha \in E$ implies that $\tau^{\alpha} \in T$. Similarly, E^{+} is a normal subgroup of E.

Proof. Let α be any element of E. An element of T is a translation τ, and τ^{α} is also a translation in T; thus $T^{\alpha} = T$. An element γ of E^{+} is a translation or a rotation whence γ^{α} is also a translation or rotation; thus $E^{+\alpha} = E^{+}$. □

REMARK. Conjugation, carrying γ to γ^{α}, can be construed as 'change of coordinates'. For example, two linear transformations A and B are similar if $B = A^{P} = P^{-1}AP$ for some invertible linear transformation (change of coordinates) P.

2. Structure of E.

We are now in a position to see how the group E is, in a certain sense, put together from groups of a simpler structure. First, we have seen that E^{+} is a normal subgroup of E. Let ρ be any element of E not in E^{+}, that is, any reflection or glide reflection. If ρ' is any other element not in E^{+}, then $\rho'\rho \in E^{+}$, whence $\rho' \in E^{+}\rho$. We have shown that E is the disjoint union of exactly two cosets of E^{+}, that is $E = E^{+} \cup E^{+}\rho$; one says that E^{+} has index 2 in E. The quotient group E/E^{+} thus has order 2, and $E/E^{+} \simeq C_2$.

Next, from the fact that τ is normal in E it follows that τ is normal in the subgroup E^{+} of E. Let 0 be any point of the plane E, and let E_0 be the stabilizer of 0 in E, that is, the set of all α in E that fix the point 0; evidently $E_0 \simeq \text{Sym}\,\Gamma$, for Γ a circle. Let γ be any element of E. There is a unique τ in T such that $0\tau = 0\gamma$. Now $\gamma\tau^{-1} = \alpha$ fixes 0, whence

$\gamma = \alpha\tau$. We say that a group G is a <u>semidirect product</u> $G = G_2G_1$ of G_1 by G_2 if G_1, G_2 are subgroups of G, G_1 normal, and $G = G_2G_1$, $G_1 \cap G_2 = 1$.

THEOREM. $E = E_0T$, <u>semidirect product of</u> T <u>by</u> E_0. <u>Similarly</u>, $E^+ = E^+{}_0T$, <u>semidirect product of</u> T <u>by</u> $E^+{}_0$.

REMARK. The semidirect product of G_1 by G_2 can equally well be written as $G = G_1G_2$ with every γ in G uniquely of the form $\gamma = \gamma_1\gamma_2$, $\gamma_1 \in G_1$, $\gamma_2 \in G_2$. (This follows from the fact that G_1 is normal in G.) This is the more natural and usual notation in a general setting, but is less suited to our geometrical interpretation.

If $G = G_2G_1 = G_1G_2$ is the semidirect product of G_1 by G_2, then the map carrying γ_2 in G_2 to the coset $G_1\gamma_2$ in G/G_1 is evidently an isomorphism from G_2 to G/G_1. Thus $E/T \simeq \mathrm{Sym}\ \Gamma$ and $E^+/T \simeq \mathrm{Sym}^+\ \Gamma$.

We note that $\mathrm{Sym}^+\ \Gamma$, the group of rotations of a circle Γ, is abelian; indeed it is isomorphic to the group (under multiplication) of all complex numbers $z = e^{i\theta}$ with $|z| = 1$. We note also that T is isomorphic to the group <u>under addition</u> of the vector space $V = V(2,\underline{R})$ of dimension 2 over the reals, whence T also is abelian. In fact, choosing a basis for V, one sees that $T \simeq \underline{R}^+ \times \underline{R}^+$, the direct product of two copies of the group $\underline{R}^+$ of real numbers under addition.

We can put all this information together as follows.

THEOREM. E <u>contains a chain of normal subgroups</u> $E > E^+ > T > 1$ <u>with successive quotients abelian, that is</u>, $E/E^+ \simeq C_2$, $E^+/T \simeq \mathrm{Sym}^+\ \Gamma$, <u>and</u> $T/1 \simeq T \simeq \underline{R}^+ \times \underline{R}^+$.

REMARK. This theorem shows that, in a certain sense, E is put together from abelian groups. However, it does not tell <u>how</u> these groups are put together. In general, knowing a normal subgroup N of a group G and knowing the quotient group G/N does not provide complete information about G. Here the details can be supplied easily, but this involves choosing a coordinate system, and we put it off to the next section.

3. Representation of $\mathcal{E}$

We choose a rectangular coordinate system in the plane E, with origin 0. Then the elements α of $\mathcal{E}_0$ are linear transformations of $V = V(2,\underline{R})$, and can be represented by matrices $M(\alpha)$. The elements σ of $\mathcal{E}^+_0$ are rotations, represented by matrices

$$M(\sigma) = \begin{pmatrix} \cos\theta & \sin\theta \\ -\sin\theta & \cos\theta \end{pmatrix},$$

where θ is the angle of σ. Reflection ρ in the x-axis is represented by

$$M(\rho) = \begin{pmatrix} 1 & 0 \\ 0 & -1 \end{pmatrix},$$

whence the element $\rho\sigma$ is represented by

$$M(\rho\sigma) = \begin{pmatrix} \cos\theta & \sin\theta \\ +\sin\theta & -\cos\theta \end{pmatrix}.$$

The image of $\mathcal{E}_0$ under this representation is the orthogonal group $O(2,\underline{R})$ consisting of all 2-by-2 real matrices M such that $MM^* = I$, where M^* is the transpose of M and $I = 1$ is the identity matrix. The image of $\mathcal{E}^+_0$ is the special orthogonal group $SO(2,\underline{R}) = O^+(2,\underline{R})$, consisting of all M in $O(2,\underline{R})$ such that $\det M = +1$.

To extend this representation to $\mathcal{E}$, that is, to accommodate translations, we need room for more parameters. We manage this by passing to 3-by-3 matrices; this apparent trick will be explained naturally when we pass to projective geometry. We pad out our coordinate system by writing $(x,y,1)$ for a point of E instead of (x,y). For an element α of $\mathcal{E}_0$, the previous matrix $M(\alpha)$ is now replaced by the matrix

$$\tilde{M}(\alpha) = \begin{pmatrix} M(\alpha) & 0 \\ 0 & 1 \end{pmatrix} = \begin{pmatrix} \cos\theta & \sin\theta & 0 \\ \mp\sin\theta & \pm\cos\theta & 0 \\ 0 & 0 & 1 \end{pmatrix}.$$

A translation $\tau \in T$, carrying (0,0) to (a,b), that is, (0,0,1) to (a,b,1), is now represented by the matrix

$$\tilde{M}(\tau) = \begin{pmatrix} 1 & 0 & 0 \\ 0 & 1 & 0 \\ a & b & 0 \end{pmatrix}.$$

We have obtained thus an isomorphism from E to the set of all matrices $\tilde{M}$ of the form

$$\tilde{M} = \begin{pmatrix} \cos\theta & \sin\theta & 0 \\ \mp\sin\theta & \pm\cos\theta & 0 \\ a & b & 1 \end{pmatrix}.$$

A simpler analytic representation of E is obtained by identifying E with the complex plane $\underline{C}$. We choose a coordinate system as before and identify the point (x,y) with the complex number x + iy. Now each translation τ is represented by a map $\tau : z \mapsto z + w$ for some $w \in \underline{C}$. Each rotation $\sigma \in E_0^+$ is represented by a map $\sigma : z \mapsto e^{i\theta}z$, for θ the angle of σ. Thus the elements α of E^+ are represented by the maps $\alpha : z \mapsto uz + w$ for $u, w \in \underline{C}$ and $|u| = 1$.

Reflection ρ in the x-axis is represented by complex conjugation, $\rho : z \mapsto \bar{z}$, where, if $z = x + iy$, $x, y \in \underline{R}$, then $\bar{z} = x - iy$. In all, E is represented by the set of all transformations of either of the two forms $z \mapsto uz + w$, $z \mapsto u\bar{z} + w$, for $u, w \in \underline{C}$ and $|u| = 1$.

4. Stabilizers and transitivity

If F is a subset of E, the <u>stabilizer</u> of F in E is the group E_F of all α in E such that $F\alpha = F$. For F a single point P we have already noted that $E_P = \text{Sym}\ \Gamma$ for any circle Γ with center P. The group $E_{P,Q} = E_P \cap E_Q$ of all transformations fixing each of two different points P and Q must fix all points of the line ℓ passing through P and Q; thus $E_{P,Q} = \{1,\rho\}$ for ρ reflection in ℓ. The stabilizer $E_{\{P,Q\}}$ of the set $\{P,Q\}$ may also contain elements interchanging P and Q, but fixing the mid-point M of $[P,Q]$. Thus $E_{\{P,Q\}} = \{1, \rho, \rho', \sigma\}$, a four group, where ρ' is reflection in the perpendicular bisector ℓ' of $[P,Q]$ and $\sigma = \rho\rho'$ is rotation about M through the angle π.

If P, Q, R are three noncollinear points, then the group fixing all three points is trivial, $E_{P,Q,R} = 1$. For any α fixing P and Q must be 1 or reflection ρ in the line $\ell = PQ$, and this ρ does not fix R. We state this as a theorem.

THEOREM. An isometry of the plane that fixes three noncollinear points is the trivial isometry.

COROLLARY. If three noncollinear points have the same images under two isometries of the plane, then the two isometries are the same.

We define two figures F_1 and F_2 in the plane E to be congruent if $F_2 = F_1\alpha$ for some isometry α in E.

THEOREM. If two triangles $[P,Q,R]$ and $[P',Q',R']$ have corresponding sides equal, then they are congruent and there is a unique isometry carrying P to P', Q to Q', and R to R'.

Proof. There is a translation τ carrying P to $P\tau = P'$. Now $d(P',Q') = d(P,Q) = d(P\tau,Q\tau) = d(P',Q\tau)$, whence Q' and $Q\tau$ are on the same circle with center P', and there is a rotation σ with center P' carrying $Q\tau$ to $Q\tau\sigma = Q'$. Thus $\tau\sigma$ carries P to P' and Q to Q'. Since $d(P',R') = d(P,R) = d(P\tau\sigma,R\tau\sigma) = d(P',R\tau\sigma)$ and $d(Q',R') = d(Q',R\tau\sigma)$, either $R' = R\tau\sigma$ or $R' = R\tau\sigma\rho$ for ρ the reflection in $\ell = P'Q'$. Thus either $\alpha = \tau\sigma$ or

$\alpha = \tau\sigma\rho$ maps P to P', Q to Q', and R to R'. Uniqueness follows from the corollary. □

REMARK. The two preceding results show why, if F is any figure in E not contained in some line, each isometry of F extends to a unique isometry of E, and Sym F is isomorphic and essentially identical with the subgroup $\mathcal{E}_F$ of $\mathcal{E}$.

The stabilizer $\mathcal{E}_\ell$ of a line ℓ will be discussed later.

5. Similarity

The group $\mathcal{E}$ plays a central role, often not explicit, in the usual development of Euclidean geometry, where one speaks of 'moving' or 'superposing' one figure F_1 on another that is congruent to it. However, the concept of distance plays only a limited role in Euclidean geometry, entering mainly in the context of the equality of two distances, or the ratio of two distances. The actual size of a geometric figure does not matter, and all its geometric properties remain unchanged if it is magnified (or reduced) by some factor, leaving all ratios of distances unchanged. Such a magnification of all distances in E is called a similarity, and we now enlarge $\mathcal{E}$ to the similarity group $\mathcal{S}$ by admitting all such transformations.

We first consider the group $\mathcal{S}_0$ of all similarities that fix the origin 0. For each real $k > 0$, let μ_k be the transformation fixing 0 and carrying each other point P to $P\mu_k = P'$ such that $\overrightarrow{0,P'} = k\cdot\overrightarrow{0,P}$; that is, μ_k is magnification by k with center 0. It is easy to see that $\mu_{k_1}\mu_{k_2} = \mu_{k_1k_2}$, whence the group M_0 of all such μ_k is isomorphic to the group $\underline{R}_+^\times$ of all positive reals under multiplication. Moreover, it is easily verified that $\mathcal{S}_0 = M_0 \times \mathcal{E}_0$, direct product, and that $\mathcal{S} = \mathcal{S}_0\mathcal{T} = M_0\mathcal{E}$, in both cases a semidirect product.

The matrix representation for E can be extended to a representation of S by representing each μ_k by the matrix

$$M(\mu_k) = \begin{pmatrix} k & 0 & 0 \\ 0 & k & 0 \\ 0 & 0 & 1 \end{pmatrix}.$$

The complex representation of E can be extended to S by the representations $\mu_k : z \mapsto kz$, $k > 0$. Since every complex number $u \neq 0$ can be written as $u = ke^{i\theta}$ for $k > 0$, the image of S contains all transformations of either of the forms $z \mapsto uz + w$ or $z \mapsto u\bar{z} + w$ for all $u, w \in \underline{C}$ with $u \neq 0$.

6. The affine group

The group E, and indeed the larger group S, have the property that they map lines to lines, but S is not the largest group with this property. The affine group A is defined to be the group of all bijections from E to E that carry lines to lines. This group might well be called the 'general linear group' if this term did not have already a different meaning; in fact, the term general linear group, $GL(2,\underline{R})$, of the real plane, is used for the stabilizer A_0 in A of a point 0.

This last assertion, that $A_0 = GL(2,\underline{R})$, is not obvious. We must show that if a bijection α of E, that maps lines to lines, fixes 0 and maps a line ℓ through 0 to itself, then α can be given by a 'linear' coordinate transformation, in the usual sense. This comes down to showing that if α fixes the points (0,0) and (1,0) of the real axis ℓ, then α fixes all points of ℓ; it is not clear that α could not permute the points (x,0) for x irrational in a nontrivial way. This will be shown in Chapter 6, but for the moment we take it for granted.

It follows, as for E and S, that $A = A_0 T$, semidirect product. Moreover, under our provisional assumption, the elements α of A_0 are represented by all nonsingular 2-by-2 real matrices

$$M(\alpha) = \begin{pmatrix} a_{11} & a_{12} \\ a_{21} & a_{22} \end{pmatrix}$$

and hence A is represented by all nonsingular 3-by-3 matrices of the form

$$\tilde{M} = \begin{pmatrix} a_{11} & a_{12} & 0 \\ a_{21} & a_{22} & 0 \\ a_{31} & a_{32} & 1 \end{pmatrix}.$$

We prove only one geometrical theorem in affine geometry.

THEOREM. *Let* [P,Q,R] *and* [P',Q',R'] *be any two triangles, that is, any two ordered triples of noncollinear points. Then there is a unique element of the affine group* A *carrying* P *to* P', Q *to* Q', *and* R *to* R'.

Proof. It suffices to prove this in the special case that $P' = 0 = (0,0)$, $Q' = X = (1,0)$, and $R' = Y = (0,1)$. There is a translation τ carrying P to $P\tau = 0$, hence carrying $[P,Q,R]$ to $[0,Q\tau,R\tau]$, where 0, $Q\tau$, and $R\tau$ are not collinear. From linear algebra we know that there is a linear transformation, in the usual sense, hence an element α of A_0, carrying X to $Q\tau$ and Y to $R\tau$. Therefore α^{-1} carries 0 to 0, $Q\tau$ to X, and $R\tau$ to Y. Thus $\tau\alpha^{-1}$ carries $[P,Q,R]$ to $[0,X,Y]$.

For uniqueness, suppose both α and β in A carried $[P,Q,R]$ to $[0,X,Y]$. Then $\alpha^{-1}\beta$ would be an element of A_0 carrying 0 to 0, X to X, and Y to Y, and therefore $\alpha^{-1}\beta = 1$, that is, $\alpha = \beta$. □

To give a very modest example of the usefulness of affine geometry, consider the theorem that the two diagonals of a parallelogram Π bisect each other. It is clear from the definition that any element α of the affine group carries parallels to parallels, and clear from the linear representation that α preserves the ratio of distances along parallel lines. Thus the assertion is true for a parallelogram Π if and only if it is true for the parallelogram $\Pi\alpha$. Now, by the theorem above, we can always choose α in such a way that $\Pi\alpha$ is a square. But the assertion is trivially true for a square. Thus it is true for all parallelograms.

The argument just given can be viewed as a generalization of the procedure common in analytic geometry where, to prove a geometric theorem, one first chooses a coordinate system judiciously.

Problems

Problem 1. When is the product of two glide reflections a rotation? When is it a translation? When is the product of four reflections a rotation? When a translation?

Problem 2. If α is an element of a group G, define the conjugation map ϕ_α from G into G by $\gamma\phi_\alpha = \gamma^\alpha$. Show that ϕ_α is an automorphism (symmetry) of G, that is, an isomorphism from G onto G. Show that the map $\Phi : \alpha \mapsto \phi_\alpha$ is a homomorphism from G into the group Aut G of all automorphisms of G. The center of G is the kernel of Φ, consisting of all elements α of G that commute with every element of G. Find the center of E, of E^+, of T.

Problem 3. The commutator subgroup (derived group) G' of a group G is the subgroup generated by all commutators $\gamma^{-1}\gamma^\alpha = \gamma^{-1}\alpha^{-1}\gamma\alpha$ of elements $\gamma,\alpha \in G$. Show that G' is a normal subgroup of G, and that, for any normal subgroup N of G, the quotient G/N is abelian if and only if N contains G'. What are the derived groups of E, E^+, T?

Problem 4. For N a normal subgroup of G, show that G is isomorphic to a semidirect product of N by G/N if and only if the standard map from G onto G/N maps some subgroup of G isomorphically onto G/N. Show that D_{2n} is isomorphic to a semidirect product of $D_{2n}^{+} \simeq C_n$ by $D_{2n}/D_{2n}^{+} \simeq C_2$.

Problem 5. If G = QN is a semidirect product of N by Q, then conjugation of elements of N by elements of Q defines a map $\phi : Q \rightarrow \text{Aut } N$. If $q_1, q_2 \in Q$ and $n_1, n_2 \in N$, then $(q_1n_1)(q_2n_2) = (q_3n_3)$ where $q_3 = q_1q_2 \in Q$ and $n_3 = (n_1(q_2\phi))n_2 \in N$. Conversely, if groups Q, N, and a map $\phi : Q \rightarrow \text{Aut } N$ are given, then the set G of all ordered pairs (q,n) for $q \in Q$, $n \in N$, becomes a group under the multiplication $(q_1,n_1)(q_2,n_2) = (q_3,n_3)$ for q_3 and n_3 as above, and G is isomorphic to a semidirect product of N by Q. G is a direct product if and only if $q\phi = 1$ for all $q \in Q$. Let $N = C_p$, $Q = C_q$, where p and q are primes; show that all semidirect products G of N by Q are isomorphic if q does not divide p - 1, but not if q does divide p - 1.

Problem 6. The general linear group $GL(2,\underline{R})$ can be defined as the automorphism group of the vector space $V = V(2,\underline{R})$. Mapping $\tau \in T$ to the vector $\overrightarrow{0,0\tau}$ defines an isomorphism from T to the additive group of V. This gives a map ϕ from E_0 to $M(E_0) \leqslant \text{Aut } T$. Show that the resulting semidirect product is isomorphic to E.

Problem 7. Let a 2-by-2 real matrix A represent a linear transformation α. Show that $A \in M(E_0) = O(2,\underline{R})$ if and only if α preserves length, and that, in that case, α also preserves magnitude of angles. Show that if T is any triangle, the area of its image $T\alpha$ is $|\det A|$ times that of T, and that, if $A \in M(E_0)$, then $\det A = \pm 1$ according as α preserves or reverses orientation.

Problem 8. Use affine geometry to prove that the three lines joining the vertices of any triangle to the midpoints of the opposite sides meet in a point.

References

This chapter should not require any background beyond that of the first chapter. The ideas of this chapter will be developed further in the next chapter. For anyone wanting a broader development, there are many books on plane Euclidean geometry; we recommend especially H. S. M. Coxeter, Introduction to Geometry, and H. W. Guggenheimer, Plane Geometry and Its Groups.

CHAPTER THREE: SUBGROUPS OF THE GROUP OF ISOMETRIES OF THE PLANE

1. Subgroups with discrete translation subgroup

THEOREM. If a subgroup G of the group $\mathcal{E}$ of all isometries of the Euclidean plane E contains no nontrivial translation, then G fixes a point.

Proof. Since G contains no nontrivial translation, it contains no glide reflection. Suppose first that G contains a nontrivial rotation σ with center 0. If α is any element of G, then σ^{α} is a rotation with center 0α, and with the same angle as σ. If $0\alpha \neq 0$, then $\sigma^{-1}\sigma^{\alpha}$ is a nontrivial translation. We conclude that every element α of G fixes 0. If G contains no nontrivial rotation, then its only nontrivial elements are reflections. Since the product of two distinct reflections is either a nontrivial translation or a nontrivial rotation, G can contain at most one reflection. Thus G is either $G = \{1\}$ or $G = \{1,\rho\}$, for ρ a reflection, and G fixes many points. □

THEOREM. If G is a finite subgroup of $\mathcal{E}$, then G is either cyclic or dihedral.

Proof. G can contain no nontrivial translation, whence G fixes a point 0. If G contains no nontrivial rotation, then $G = \{1\}$ or $G = \{1,\rho\}$ as before, of type C_1 or D_2. Otherwise, let G contain a rotation σ of least positive angle θ. Clearly G cannot contain a rotation σ_1 with angle θ_1 such that $n\theta < \theta_1 < (n+1)\theta$ for any $n \in \underline{Z}$. Thus G^+ is cyclic with generator σ_1, and G is either G^+ or dihedral. □

2. Frieze groups

We now consider subgroups G of $\mathcal{E}$ such that the translation subgroup $T = G \cap \mathcal{T}$ is infinite cyclic. Such groups are the symmetry groups of certain infinite plane figures admitting, as translational symmetries, only iterates (powers) of some translation along an axis ℓ. Such figures are called friezes, and their symmetry groups are the frieze groups. We shall enumerate the types of frieze groups, illustrating them by giving, for each type, a frieze whose symmetry group is of that type.

Let F be a frieze group, and let τ be a generator for T.

Suppose first that F contains no nontrivial rotation. Then $F^+ = F \cap \mathcal{E}^+ = T$. It may be that $F = T$. Otherwise $F = \langle \tau, \rho \rangle$, the group generated by τ and ρ, where ρ is a reflection or a glide reflection. Since $T^\rho = T$, and τ generates T, τ^ρ also must generate T, hence either $\tau^\rho = \tau$ or $\tau^\rho = \tau^{-1}$, and ρ has axis parallel or perpendicular to an axis ℓ of τ. We cannot have elements ρ of both kinds, for then their product would be a nontrivial rotation. If ρ is a reflection, it can be of either kind. If ρ is a glide reflection, then ρ^2 is a nontrivial translation, hence $\rho^2 = \tau^h$ for some $h \neq 0$, and ρ has axis ℓ. Now ρ commutes with τ, and $(\rho\tau^k)^2 = \tau^{h+2k}$. Replacing ρ by $\rho\tau^k$ for suitable k we may suppose that either $\rho^2 = 1$ and ρ is a reflection, or that $\rho^2 = \tau$ and ρ is a glide reflection; in the latter case F contains no reflection.

We have obtained four geometrically different types, as follows:

$$F_1 = \langle \tau : \emptyset \rangle, \text{ of isomorphism type } C_\infty;$$
$$F_1^1 = \langle \tau, \rho : \rho^2 = 1,\ \tau^\rho = \tau \rangle, \text{ isomorphic to } C_\infty \times C_2;$$
$$F_1^2 = \langle \tau, \rho : \rho^2 = 1,\ \tau^\rho = \tau^{-1} \rangle, \text{ isomorphic to } D_\infty;$$
$$F_1^3 = \langle \tau, \rho : \rho^2 = \tau,\ \tau^\rho = \tau \rangle, \text{ isomorphic to } C_\infty.$$

We emphasize that F_1 and F_1^3 are isomorphic as abstract groups but are not geometrically equivalent (that is, as subgroups of $\mathcal{E}$), since F_1 preserves orientation while F_1^3 does not.

There remains only the case that F contains a nontrivial rotation σ. Again, τ^σ must generate $T^\sigma = T$, and hence $\tau^\sigma = \tau$ or $\tau^\sigma = \tau^{-1}$. Since $\sigma \neq 1$, we cannot have $\tau^\sigma = \tau$. Thus $\tau^\sigma = \tau^{-1}$, and σ is a rotation of order 2. We may now choose the axis ℓ of τ through the center 0 of σ. If ρ' is any other rotation in F, then it also must have order 2, whence $\sigma\sigma'$ is a translation, $\sigma\sigma' = \tau^h$ for some h, and $\sigma' = \sigma\tau^h$. Thus $F^+ = < \tau,\sigma : \sigma^2 = 1, \tau^\sigma = \tau^{-1} >$.

It may be that $F = F^+$. Otherwise $F = < \tau,\sigma,\rho >$, generated by τ, σ, and ρ where ρ is a reflection or glide reflection. As before, the axis ρ must be either parallel or perpendicular to ℓ. If ρ has axis perpendicular to ℓ, then $\rho' = \rho\sigma$ has axis parallel to ℓ. Thus, replacing ρ by ρ' if necessary, we may suppose that ρ has axis ℓ' parallel to ℓ. If $\ell' \neq \ell$, then $0\rho \neq 0$, and the line $m = \overline{0,0\rho}$ would be perpendicular to ℓ. Now the product of σ and σ^ρ, rotations of order 2 with centers 0 and 0ρ, would be a translation in the direction of m, perpendicular to ℓ, contrary to the hypothesis that T is infinite cyclic with generator τ. Thus ρ has axis ℓ.

If ρ is a reflection, then $\rho_1 = \rho\sigma$ is also a reflection, with axis m the perpendicular to ℓ at 0.

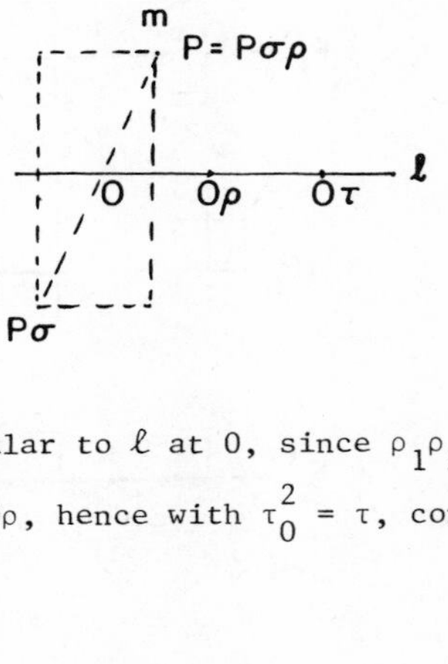

If ρ is a glide reflection, we may suppose as before that $\rho^2 = \tau$. Let m be the perpendicular bisector of $[0,0\rho]$. Inspection of the figure shows that $\rho_2 = \sigma\rho$ fixes all points of m, hence ρ_2 is the reflection in m. In this case F contains no reflection ρ_1 with axis perpendicular to ℓ at 0, since $\rho_1\rho_2$ would then be a translation τ_0 carrying 0 to 0ρ, hence with $\tau_0^2 = \tau$, contrary to the assumption that τ generates T.

We have obtained three geometrically different types of groups F containing a nontrivial rotation, as follows:

$$F_2 = \langle \tau,\sigma : \sigma^2 = 1,\ \tau^\sigma = \tau^{-1} \rangle, \text{ isomorphic to } D_\infty;$$

$$F_2^1 = \langle \tau,\sigma,\rho : \sigma^2 = 1,\ \tau^\sigma = \tau^{-1},\ \rho^2 = 1,\ \tau^\rho = \tau,\ \sigma^\rho = \sigma \rangle, \text{ isomorphic to } D_\infty \times C_2;$$

$$F_2^2 = \langle \tau,\sigma,\rho : \sigma^2 = 1,\ \tau^\sigma = \tau^{-1},\ \rho^2 = 1,\ \tau^\rho = \tau^{-1},\ \sigma^\rho = \sigma\tau \rangle,$$
$$= \langle \sigma,\rho : \sigma^2 = 1,\ \rho^2 = 1 \rangle, \quad \text{isomorphic to } D_\infty \simeq C_2 \star C_2.$$

We note again that the abstractly isomorphic types F_2 and F_2^2 are geometrically different, since F_2 preserves orientation while F_2^2 does not.

THEOREM. _There are exactly seven geometrically distinct types of frieze groups, with presentations as given above. They fall into the four isomorphism types of_ C_∞, D_∞, $C_\infty \times C_2$, $D_\infty \times C_2$.

In the following figure we show friezes with symmetry groups of the seven types. In these figures, τ is a horizontal translation. Broken lines indicate axes of reflections or glide reflections. In the figures for F_2, F_2^1, and F_2^2, small circles mark centers of rotational symmetry.

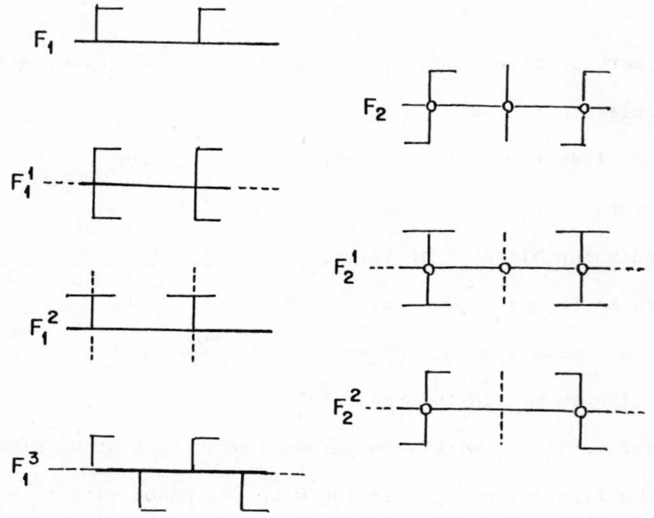

F_1 F_2 F_1^1 F_2^1 F_1^2 F_2^2 F_1^3

3. Discontinuous groups

All the frieze groups leave invariant a line ℓ, the axis of the frieze, whence they are all contained in the stabilizer E_ℓ of some line ℓ. An argument that we have used repeatedly gives the structure of E_ℓ.

THEOREM. The stabilizer E_ℓ of a line ℓ in E is a semidirect product $E_\ell = E_{\ell,0} \cdot T_\ell$, where $E_{\ell,0} \simeq D_4$ is the stabilizer in E_ℓ of a point 0 on ℓ, and $T_\ell = E_\ell \cap T$, the stabilizer of ℓ in T, is isomorphic to $\underline{R}^+$, the group of reals under addition.

The group E_ℓ contains a great variety of subgroups, many of which are of little geometrical interest. For example, the group of all translations through distances of the form $a + b\sqrt[3]{7}$ for $a, b \in \underline{Q}$. Geometrically interesting subgroups of E are usually very 'large' in some sense, and 'continuous', like E_ℓ, or are rather sparse, like finite subgroups or the frieze groups, and are 'discontinuous'.

DEFINITION. A subgroup G of E is discontinuous if, for every point P of the plane E, there is some disc D with center P that contains no image $P\alpha$, $\alpha \in G$, of P other than P itself.

We note that this condition is equivalent to the following: if P is any point of E and PG is the orbit $PG = \{P\alpha : \alpha \in G\}$, and if D is any disc in E, then the intersection of PG with D consists of only finitely many points.

We note also that if G is a discontinuous subgroup of E, then every subgroup H of G is also discontinuous, in particular $T = G \cap T$ is discontinuous.

THEOREM. If T is a discontinuous subgroup of T, then T is trivial, infinite cyclic, or free abelian of rank 2, that is, $T = 1$, $T \simeq C_\infty$, or $T \simeq C_\infty \times C_\infty$.

Proof. Let T be a discontinuous subgroup of $\mathcal{T}$. Assume that $T \neq 1$. Let $\tau \in T$, $\tau \neq 1$, and let ℓ be a line, through a point 0, that is invariant under τ. Let T_ℓ be the stabilizer of ℓ in T. If D is any disc with center 0 and containing 0τ, then D contains only finitely many points 0α for $\alpha \in T_1$, and, since $0\tau \neq 0$, it contains some such $0\alpha \neq 0$ at a least distance from 0. We will show that T_1 is infinite cyclic with generator α. If τ is any element of T_1, then 0τ lies on ℓ, hence in the closed interval $[0\alpha^k, 0\alpha^{k+1}]$ for some k in $\underline{Z}$. If $0\tau = 0\alpha^k$ or $0\tau = 0\alpha^{k+1}$, then $\tau = \alpha^k$ or $\tau = \alpha^{k+1}$. Otherwise $0 < d = d(0\alpha^k, 0\tau) < d(0, 0\alpha)$, and $\tau' = \alpha^{-k}\tau$, which translates $0\alpha^k$ to 0τ, through a distance d, translates 0 to $0\tau'$, at a distance $d < d(0, 0\alpha)$ from 0, contrary to the choice of α.

If $T = T_1$, then T is infinite cyclic. We suppose now that T is not infinite cyclic. We now revise our choice of $T_1 = \langle \alpha \rangle$, taking $\alpha \neq 1$ such that $a = d(0, 0\alpha) > 0$ is minimal. Since $T \neq T_1$, we can choose $\beta \in T$, with $\beta \notin T_1$, such that $b = d(0, 0\beta) \geqslant a$ is minimal. Let Π be the closed parallelogram with vertices 0, 0α, 0β, and $0\alpha\beta$. It is clear that the images $\Pi\gamma$ for all $\gamma = \alpha^m\beta^n \in T_2$ fill out the plane E. Let τ be any element of T. Then 0τ lies in some $\Pi\gamma$, $\gamma \in T_2$, whence $0\tau\gamma^{-1}$ lies in Π. To show that $\tau \in T_2$ it suffices to show that $\tau' = \tau\gamma^{-1} \in T_2$. If $0\tau'$, in Π, is one of the vertices of Π, then τ' is one of 1, α, β, $\alpha\beta$, and $\tau' \in T_2$. If $0\tau'$ lay on one of the sides $[0, 0\alpha]$ or $[0\beta, 0\alpha\beta]$ of Π, then $0\tau'$ or $0\beta^{-1}\tau'$ would be nearer to 0 than 0α, contrary to the choice of α. Thus we can suppose that $0\tau'$ is not on either of these sides. If $0\tau'$ were in the open disc D at 0 with radius b, it would be nearer to 0 than 0β, contrary to the choice of β. If $0\tau'$ were in the open disc $D\alpha\beta$ at $0\alpha\beta$ with radius b, then

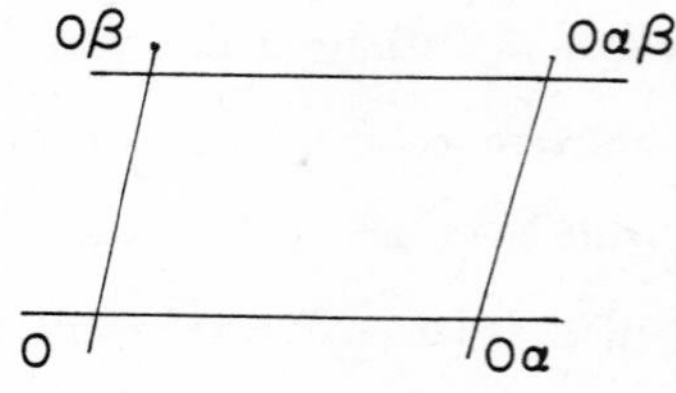

$0\alpha^{-1}\beta^{-1}\tau'$ would be nearer to 0 than 0β, again contrary to the choice of β. But, since $a \leqslant b$, the union of the discs D and $D\alpha\beta$ includes all of Π except possibly the vertices 0α, 0β. This is a contradiction, completing the proof that $T = T_2 = \langle \alpha, \beta \rangle$, free abelian of rank 2. □

THEOREM. _If_ G _is a discontinuous subgroup of_ $\mathcal{E}$, _then the stabilizer_ G_0 _in_ G _of a point_ 0 _is finite, either cyclic or dihedral_.

Proof. It was seen earlier that $\mathcal{E}_0 = \mathrm{Sym}\,\Gamma$ for any circle Γ with center 0. Thus G_0 is a subgroup of $\mathrm{Sym}\,\Gamma$. Let P be any point on Γ and D any disc containing Γ. Since the orbit $PG_0 \subseteq \Gamma \subseteq D$, and G_0 is discontinuous, PG_0 is a finite set of points on Γ. Since PG_0 is invariant under G_0, its points are equally spaced, as the vertices of a regular polygon F. Now G_0, permuting these points, is contained in the dihedral group Sym F. □

THEOREM. _If_ G _is a discontinuous subgroup of_ $\mathcal{E}$ _and_ $T = G \cap \mathcal{T}$ _is trivial, then_ G _is a finite group, either cyclic or dihedral_.

Proof. We have seen that if G contains no nontrivial translation, then it fixes some point 0, that is, $G = G_0$, whence the preceding theorem applies. □

THEOREM. _If_ G _is a discontinuous subgroup of_ $\mathcal{E}$ _and_ $T = G \cap \mathcal{T}$ _is infinite cyclic, then_ G _is a frieze group_.

Proof. This was the definition of a frieze group. □

The case remains that G is a discontinuous subgroup of $\mathcal{E}$ and $T = G \cap \mathcal{T}$ is free abelian of rank 2. These groups are called (planar) _crystallographic groups_, and are of considerable interest both in mathematics and its applications. However, an exhaustive enumeration of them is moderately difficult, and we put it off to the next chapter. For the present we examine only the easiest and perhaps most important of them, introducing methods for which we shall find repeated application.

4. Regular tessellations of the Euclidean plane

A tessellation (tiling) of the plane E is a division of E into non-overlapping closed regions, which we shall always assume are bounded by finite polygons. A polygon with n sides, $n \geqslant 3$, is regular if its sides are of equal length and the n interior angles are all equal. A tessellation is regular if all the faces (tiles) into which it divides the plane are bounded by congruent regular polygons.

It turns out that there are only three types, under similarity, of regular tessellations F. For each of these we will determine its symmetry group Sym F.

Let F be a regular tessellation by polygons Π with $n \geqslant 3$ sides. Since all the interior angles are equal, there must be the same number $m \geqslant 3$ of such polygons at each vertex. As we go around the polygon Π, at each vertex we turn through an angle of $2\pi/n$, whence the interior angles must be $\pi - 2\pi/n$. Since there are m polygons Π at each vertex, the interior angle in each must be $2\pi/m$. We conclude that $\pi - \frac{2\pi}{n} = \frac{2\pi}{m}$, or $\frac{1}{n} + \frac{1}{m} = \frac{1}{2}$. We cannot have $m,n \geqslant 5$, whence one of m, n must be 3 or 4. One checks that there are exactly three solutions: $(n,m) = (3,6), (4,4), (6,3)$. These describe the regular tessellations by equilateral triangles, six at each vertex, by squares, four at each vertex, and by regular hexagons, three at each vertex.

To determine the structure of the symmetry groups G = Sym F of these tessellations we could proceed as with E, obtaining a description of G in terms of its translation subgroup and of stabilizers of points. However, we prefer to introduce an important method of Poincaré, by which one obtains a presentation of G from geometrical considerations. This method will be presented in detail in Chapter 10, and here we present it less formally, with substantial appeal to geometric intuition. With this purpose, we confine attention to the tessellation by squares, the most easily visualized; the treatment of the other two cases follows exactly the same pattern.

Our method rests upon finding a <u>fundamental region</u> for G, that is, a polygonal region Δ whose images Δg under elements $g \in G$ are distinct, for distinct $g \in G$, and fill out the plane without overlapping. Let Q be one of the square tiles of the tessellation F. Evidently $G = G_Q T$, for $T = G \cap \tau$, where $G_Q \cap T = 1$. It follows that we may choose Δ as a fundamental region for the action of G_Q on Q. Now $G_Q = \text{Sym } Q \simeq D_8$, and it is clear that we can take Δ as shown in the figure, the triangular part of Q cut out by successive axes of reflections.

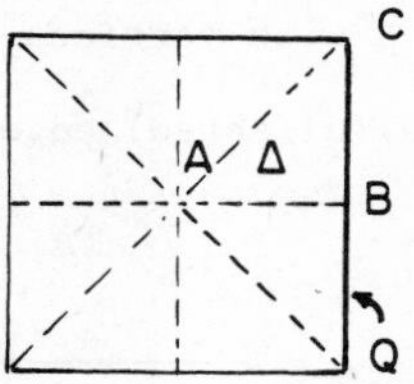

Let A, B, C be the vertices of Δ, as shown, and let α, β, γ be reflections in the sides a, b, c of Δ opposite A, B, C. Clearly $\alpha, \beta, \gamma \in G$.

THEOREM. G <u>is generated by</u> α, β, <u>and</u> γ.

<u>Proof</u>. Let H be the subgroup of G generated by α, β, γ. We must show that H = G. Suppose not. Since E is the nonoverlapping union of the Δg, $g \in G$, with the Δg distinct for different g, the union U of the Δh for $h \in H$ cannot be all of E. Thus some side s of some triangle Δh, $h \in H$, must lie on the boundary of U, separating Δh from some Δg, $g \notin H$. Now s is the image ah, bh, or ch of a side a, b, c of Δ, and we have seen that reflection ρ in s is the corresponding element α^h, β^h, or γ^h. Since $h, \alpha, \beta, \gamma \in H$, ρ is in H, and $g = h\rho \in H$, a contradiction. □

We now seek a set of defining relations among α, β, and γ. Since α, β, γ are reflections, we have relations $\alpha^2 = 1$, $\beta^2 = 1$, $\gamma^2 = 1$. Further, $\beta\gamma$ is a rotation about A through an angle twice the interior angle $2\pi/8$ of Δ at A, that is, through $2\pi/4$; therefore, $(\beta\gamma)^4 = 1$. Similarly, $(\gamma\alpha)^2 = 1$ and $(\alpha\beta)^4 = 1$. We shall show that there are a full set of defining relations.

THEOREM. G <u>has a presentation</u>

$$G = \langle \alpha,\beta,\gamma : \alpha^2,\beta^2,\gamma^2,(\beta\gamma)^4,(\gamma\alpha)^2,(\alpha\beta)^4 \rangle.$$

<u>Proof</u>. Let D be the tessellation (not a regular tessellation) of E by the regions Δg, $g \in G$. Let 0 be the center (intersection of lines joining vertices to midpoints of opposite sides) of Δ. We construct a graph C with vertices all centers $0g$ of triangles Δg. We join $0g$ and $0h$ with a (directed) edge e if and only if Δg and Δh have a side s in common.

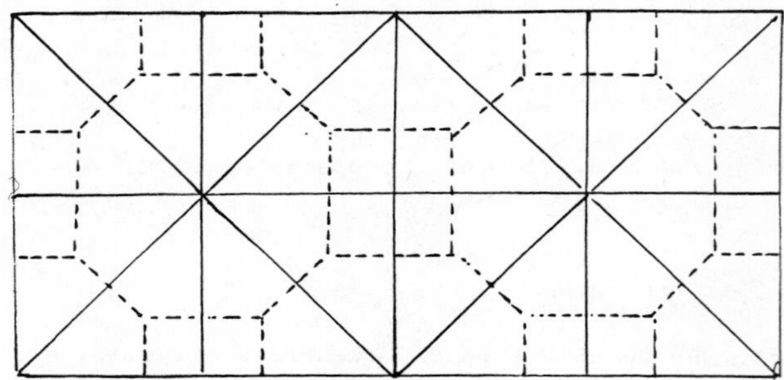

We assign to e the <u>label</u> $\lambda(e) = \alpha,\beta,\gamma$, according as s is the image ag, bg, cg of the side a, b, c of Δ. (The graph C with this label function is an example of a <u>Cayley graph</u>.)

Let $p = e_1 \ldots e_n$, $n \geqslant 0$, be a <u>path</u> in C, that is a sequence of edges such that, for $0 \leqslant i < n$, e_{i+1} begins where e_i ends. Let p, beginning at some $0g_0$, have successive vertices $0g_0, 0g_1, \ldots, 0g_n$. We define $\lambda(p) = x_1 \ldots x_n$ where $x_i = \lambda(e_i) = \alpha,\beta$, or γ. It was seen in the previous proof that reflection ρ in the edge s_i between Δg_i and Δg_{i+1} is $\rho = x_i^{g_i}$, whence $g_{i+1} = g_i\rho = g_i x_i^{g_i} = x_i g_i$. It follows that $g_n = x_n \ldots x_1 g_0$. We define $\lambda^*(p) = x_n \ldots x_1$; then $g_n = \lambda^*(p) g_0$.

Note that if p is a closed path (beginning and ending at the same point $0g_0 = 0g_n$) then $g_0 = g_n$, and the relation $\lambda^*(p) = 1$ holds in G. Conversely, if $w = x_n \ldots x_1$ and $w = 1$ is a relation in G, and g_0 is arbitrary, there is a unique closed path p at $0g_0$ such that $\lambda^*(p) = w$.

The graph C divides the plane up into regions Δ^*, some square and some hexagonal, thus defining a tessellation D^* (dual to D). Suppose that a path p runs between points P and Q along an arc on one side of the boundary of a region Δ^*, and that p' is obtained from p by replacing this arc by the arc from P to Q running around the other side of Δ^*. The label on a path running once around the boundary of Δ^* is easily seen to be one of $(\beta\gamma)^{\pm 4}$, $(\gamma\alpha)^{\pm 2}$, $(\alpha\beta)^{\pm 4}$, whence it is easily seen that the relation $\lambda^*(p) = \lambda^*(p')$ follows from the corresponding relation $(\beta\gamma)^4 = 1$, $(\gamma\alpha)^2 = 1$, or $(\alpha\beta)^4 = 1$.

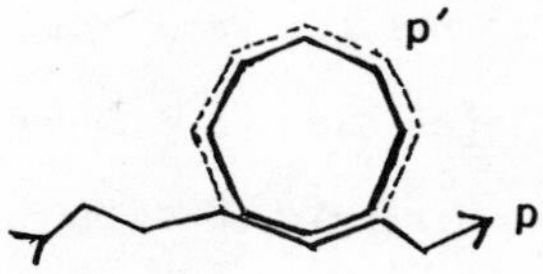

p'
p

The relations $\alpha^2 = 1$, $\beta^2 = 1$, $\gamma^2 = 1$ correspond similarly to modifying p by deleting (or inserting) a 'spine', that is an edge e_i followed by an edge e_{i+1} that is the same edge except traversed in the opposite direction.

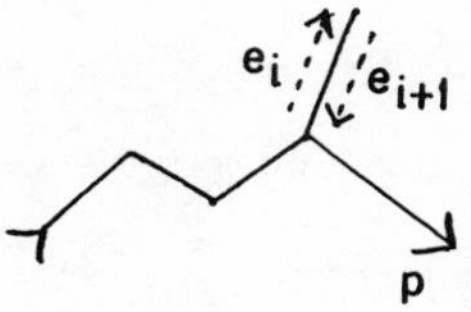

e_i
e_{i+1}
p

To show that the given relations define G is therefore equivalent to showing that any closed path in C can be reduced to the trivial path at some point (with n = 0 edges) by a succession of modifications of these two kinds. This is fairly obvious intuitively (one can think of contracting p while successively lifting it over pegs at the centers of the regions Δ^*), and is easily proved. The proof reduces to the case that p is a simple loop (does not intersect itself), where, inductively, by running around the other side of some region Δ^*, one can reduce the number of regions Δ^* enclosed by ρ. □

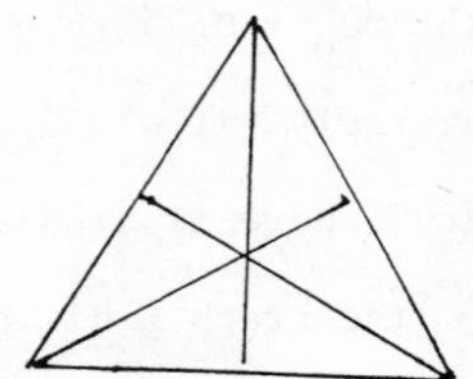

For the tessellation F of type (3,6), by equilateral triangular regions, six at each vertex, the argument differs from the case above only in that the interior angles of Δ are different. We find a presentation

$$G = \langle \alpha,\beta,\gamma : \alpha^2 = \beta^2 = \gamma^2 = (\beta\gamma)^3 = (\gamma\alpha)^2 = (\alpha\beta)^6 = 1 \rangle.$$

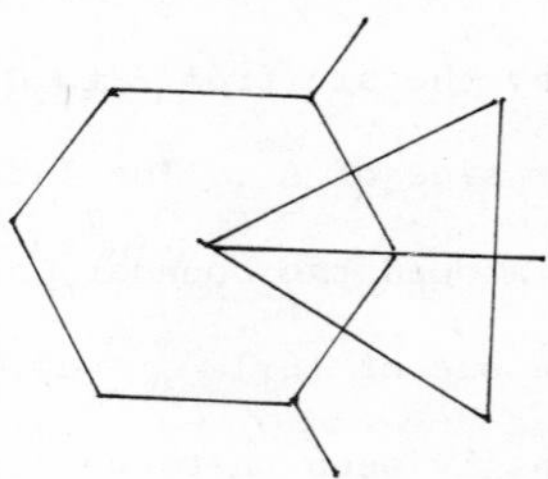

The remaining tessellation, of type (6,3) is dual to one of type (3,6), and hence has the same symmetry group G; this can also be deduced from the fact that a single fundamental region Δ serves for both of them.

We omit the proof, which is elementary but slightly tricky, that if G is a group with presentation

$$G = \langle \alpha,\beta,\gamma : \alpha^2 = \beta^2 = \gamma^2 = (\beta\gamma)^p = (\gamma\alpha)^q = (\alpha\beta)^r = 1 \rangle,$$

then the subgroup G^+ generated by $x = \beta\gamma$, $y = \gamma\alpha$, and $z = \alpha\beta$, has a presentation

$$G^+ = \langle x,y,z : x^p = y^q = z^r = xyz = 1 \rangle,$$

or, equivalently,

$$G^+ = \langle x,y : x^p = y^q = (xy)^r = 1 \rangle.$$

5. Triangle groups

The groups studied above are special <u>Coxeter groups</u>, that is, discontinuous groups in Euclidean (or other) space of dimension $n \geqslant 2$ generated by reflections ρ_i. The product $\rho_i\rho_j$ of two of these reflections, if it has finite order, must be a rotation through an angle $2\pi/m_{ij}$ for some positive integer m_{ij}. The group G will have a presentation with generators ρ_i and defining relations $(\rho_i\rho_j)^{m_{ij}} = 1$, where $m_{ij} = m_{ji}$ and all $m_{ii} = 1$.

A <u>triangle group</u> is a discontinuous group G generated by reflections α, β, γ in the three sides of a triangle, bounding a fundamental region Δ for G. The interior angles must be $\frac{\pi}{p}$, $\frac{\pi}{q}$, $\frac{\pi}{r}$ for integers $p, q, r \geqslant 2$. In Euclidean plane geometry the sum of the interior angles must be π, whence we must have $\frac{1}{p} + \frac{1}{q} + \frac{1}{r} = 1$. We have encountered already the cases where (p,q,r) are, in some order, (4,4,2) and (3,6,2). There remains only one other case, (3,3,3), where Δ is bounded by an equilateral triangle, and the group is

$$G = \langle \alpha,\beta,\gamma : \alpha^2 = \beta^2 = \gamma^2 = (\beta\gamma)^3 = (\gamma\alpha)^3 = (\alpha\beta)^3 = 1 \rangle.$$

This group is evidently a subgroup of the full symmetry group of the regular tessellation of E by equilateral triangles.

Triangle groups with $\frac{1}{p} + \frac{1}{q} + \frac{1}{r} > 1$ are not possible in the Euclidean plane, but they are realizable on the sphere S, by taking Δ bounded by a spherical triangle [A,B,C], whose sides a, b, c are arcs of great circles, meeting with interior angles π/p, π/q, π/r, whose sum is <u>greater than</u> π. The case is entirely analogous to that in the plane except for the fact that, since the sphere S has finite area, the resulting tessellation has only finitely many faces Δg, and hence the group G is finite.

Instead of working on the sphere S, one can replace the sides a, b, c of the spherical triangle by the planes $\bar{a}$, $\bar{b}$, $\bar{c}$ in Euclidean three dimensional space E^3, passing through the center 0 of S and containing the arcs a, b, c. One can then replace the reflections α, β, γ by the reflections $\bar{\alpha}$, $\bar{\beta}$, $\bar{\gamma}$ of E^3 in the planes $\bar{a}$, $\bar{b}$, $\bar{c}$. These generate a group $\bar{G}$ isomorphic to G. In fact, $\bar{G}$ obviously leaves S invariant, and G is obtained from $\bar{G}$ by restricting each element $\bar{g}$ of $\bar{G}$ to the transformation g that it effects on S.

6. Regular solids in three dimensional Euclidean space

A regular tessellation F of S, of type (n,m), will have faces that are congruent regular spherical n-gons, with m meeting at each vertex. It will have only finitely many faces. If Π is any face in F, then all of the vertices of Π lie in some plane p, and determine a regular n-gon Π' in the plane p. The region of E^3 bounded by all the Π' is a regular solid, bounded by congruent regular planar n-gons, with m at each vertex. Conversely, given any regular solid, projection of its faces onto a circumscribed sphere S gives a regular tessellation F of S.

To find the symmetry group G = Sym F of such a tessellation we proceed exactly as in the Euclidean case, choosing as fundamental region one of the 2n triangular regions Δ into which a face Π is divided by the axes of its reflectional symmetries. One of the angles of Δ, say at C, will be $\pi/2$, and the others will be π/n, π/m, where $1/n + 1/m + 1/2 > 1$, or $1/n + 1/m > 1/2$. (We note that the condition $1/p + 1/q + 1/r > 1$ in fact implies that at least one of p, q, r be 2.)

Case 0. On the sphere it is technically possible to have one of n, m = 2, and the other an arbitrary integer greater than or equal to 2. If n = 2, then Π is a 2-gon (digon), whose two sides are halves of great circles meeting at two antipodal vertices, say the north pole N and south pole S. If the interior angle is $2\pi/m$, there will be m of these digons meeting at each of the two vertices N and S. The group G contains, as subgroup of index 2, the group

$G^+ = \langle x,y,z : x^2 = y^m = z^2 = xyz = 1 \rangle = \langle x,z : x^2 = z^2 = (zx)^m = 1 \rangle$,

dihedral of order 2m. The group G^+ in turn contains a cyclic group $H \simeq C_m$ of index 2 in G^+, hence of index 4 in G, with Π as fundamental region.

If m = 2, then Π is an n-gonal region with all interior 'angles' π, that is, Π is a hemisphere, bounded by an equator divided into n equal arcs. There are only two faces, Π and a complementary hemisphere with the same sides and vertices. The symmetry group G is the same.

Note that these degenerate tessellations of the sphere do not give rise to any regular solid in the usual sense.

We suppose henceforth that n, m $\geqslant$ 3. The condition $1/n + 1/m > 1/2$ requires that one of n, m be 3. We suppose that m = 3, and return to the dual cases with n = 3 later. Now we must have $1/n > 1/6$, whence n = 3, 4, or 5.

<u>Case 1</u>: (n,m) = (3,3). The faces Π of F are equilateral triangular regions, with three at each vertex. One sees that F is obtainable by projecting a regular <u>tetrahedron</u> onto a circumscribed sphere S.

Since F has four faces Π, each made up of $2 \cdot 3 = 6$ fundamental regions Δg, there are in all $6 \cdot 4 = 24$ copies Δg of Δ, whence G has order 24. Alternatively, one can see that G effects all permutations of the set of the four vertices, whence G is isomorphic to S_4, the symmetric group of all permutations of four objects. The subgroup G^+ corresponds thus to A_4, the alternating group, of index 2 in S_4, hence of order 12.

<u>Case 2</u>: (n,m) = (4,3). Here F is the projection of the <u>cube</u> onto a circumscribed sphere S. The cube has 6 faces Π, each made up of 8 copies of Δ, whence G has order 48.

Case 3 : $(n,m) = (5,3)$. Here F is the projection of the regular dodecahedron, with 12 pentagonal faces, three at each vertex. Since this solid is less familiar than the tetrahedron and cube, and also to illustrate a method that will be used later in a more complicated context, we indicate how to construct F, or equivalently, how to construct the regular dodecahedron, that is, how to prove that it exists.

To construct F we need a regular spherical pentagon Π with interior angles $2\pi/3$. Let Π be a regular spherical pentagon with radius (distance from its center to its vertices) r in the range $0 < r < r_1$ where r_1 is the length of one quarter of a great circle. As r approaches 0, Π becomes more nearly Euclidean, and its interior angle θ approaches $\pi - \frac{2\pi}{5} = \frac{3\pi}{5}$. As r approaches r_1, Π approaches a great circle divided into five equal arcs as sides, and θ approaches π. Since θ increases continuously with r, and $\frac{3\pi}{5} < \frac{2\pi}{3} < \pi$, for some value of r the pentagon Π has interior angle $\theta = \frac{2\pi}{3}$.

Now three copies of Π will fit together at a vertex, each pair having a side in common. We start with a central face Π_1 (shown shaded in our rough sketch) and adjoin the remaining faces at its vertices, forming a (unshaded) ring R of five faces enclosing Π_1. Five more faces with sides in common with those of R form a second (shaded) ring R', enclosing the previous figure. The boundary of the resulting figure is now a cycle of five sides, and the complementary region (out of sight in the sketch) can be filled in with a twelfth pentagon.

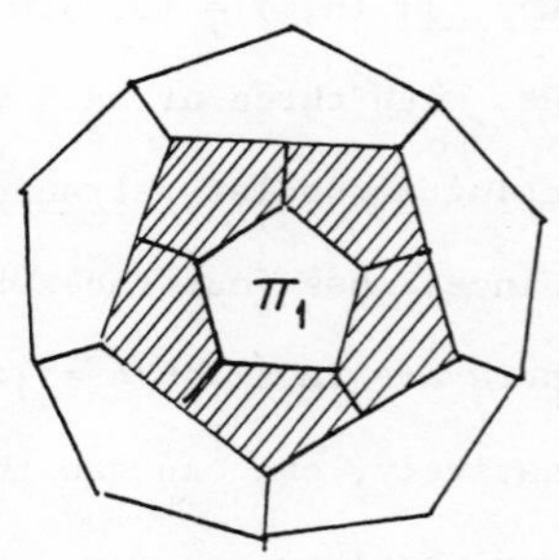

Alternatively, one can construct the dodecahedron in E^3 by starting with 12 congruent regular pentagons, say cut out of pasteboard. Hinge two of them together along a common side, and bend at the hinge until the angle is right for a third to be fitted at one end of the hinge, with a side in common with each of the others. This forms a rigid figure, and the angles between each pair of faces are equal and just right to permit attachment of more faces until the figure closes up with attachment of the twelfth face, as before.

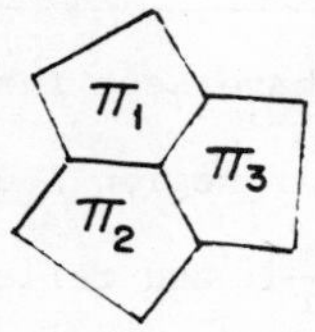

π_1 π_2 π_3

We note also that, leaving out the metric arguments, which appeal to continuity, we have proved the existence of a combinatorial tessellation F of the sphere S, possibly irregular, by faces each abutting on exactly five others, and with exactly three meeting at each vertex. Moreover, F is combinatorially unique, that is, unique up to homeomorphism.

The group G has order $10 \cdot 12 = 120$.

The dual cases obtained by interchanging n and m were passed over above. As in the plane case, the dual tessellation F^* has the same symmetry group G as the tessellation F. In the case (3,3) of the tetrahedron, F^* is of the same type as F, and the dual regular solid obtained from F^* is again a tetrahedron. For the cube (4,3), we obtain as dual solid the regular octahedron (3,4); it has 8 triangular faces (corresponding to the 8 vertices of the cube), with 4 meeting at each of its 6 vertices (corresponding to the 6 faces of the cube). For the dodecahedron (5,3), we obtain as its dual the regular icosahedron (3,5), with 20 triangular faces, 5 meeting at each of its 12 vertices. There are in all exactly five types of (nondegenerate) regular solids.

In Chapter 5 we will examine the regular 'hypersolids' in Euclidean space of dimension greater than three.

7. Hyperbolic triangle groups

We have seen that if $\frac{1}{p} + \frac{1}{q} + \frac{1}{r} = 1$, for integers p, g, r $\geqslant 2$, a triangular region Δ exists in the Euclidean plane E, with interior angles $\frac{\pi}{p}, \frac{\pi}{q}, \frac{\pi}{r}$, and reflections in the three sides of Δ generate a discontinuous group G with Δ as fundamental region.

We have seen also that if $\frac{1}{p} + \frac{1}{q} + \frac{1}{r} > 1$, there exists a triangular region Δ on the sphere S with angles $\frac{\pi}{p}, \frac{\pi}{q}, \frac{\pi}{r}$, and reflections in the three sides of Δ generate a discontinuous, in fact, finite group G with Δ as fundamental region.

If $\frac{1}{p} + \frac{1}{q} + \frac{1}{r} < 1$, the situation is analogous in the <u>hyperbolic plane</u> H. This will be discussed in some detail in Chapters 9 and 10, and we say no more here than to describe one example (omitting all formal details).

We describe a tessellation F of the hyperbolic plane H of type (5,4), by regular pentagons, four at each vertex. The hyperbolic plane H can be described as the open unit disc, with a hyperbolic distance function different from the Euclidean distance, which we need not define here. The <u>hyperbolic lines</u> in H are the intersections with H of Euclidean circles (and lines) orthogonal to the unit circle. It is not difficult to see that one can find five circles $C_1, \ldots, C_5$ of

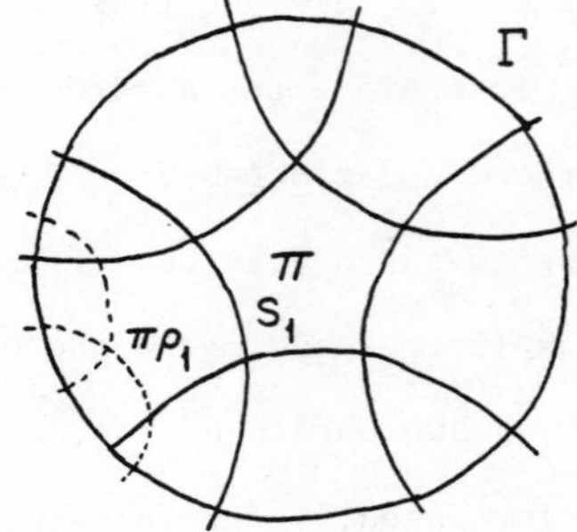

equal radius and with centers equidistant from the center O of and equally spaced radially, all orthogonal to Γ and such that arcs $s_1,\ldots,s_5$ of $C_1,\ldots,C_5$ form the five sides of a rectangular (hyperbolic) pentagon Π.

Hyperbolic reflection ρ_i in the side s_i of Π is understood as inversion in the circle C_i (see Chapter 9), which leaves H invariant. Now Π is a fundamental region for the group G generated by $\rho_1,\ldots,\rho_5$, and the set of images Πg for all $g \in G$ is the set of faces of a regular tessellation F of H by rectangular pentagons.

The group G has a presentation

$$G = < \rho_1,\ldots,\rho_5 : \rho_1^2 = \ldots = \rho_5^2 = (\rho_1\rho_2)^2 = \ldots = (\rho_5\rho_1)^2 = 1 >,$$

and the subgroup G^+, generated by $\sigma_1 = \rho_1\rho_2,\ldots,\sigma_5 = \rho_5\rho_1$, has a presentation

$$G^+ = < \sigma_1,\ldots,\sigma_5 : \sigma_1^2 = \sigma_2^2 = \sigma_3^2 = \sigma_4^2 = \sigma_5^2 = \sigma_1\sigma_2\sigma_3\sigma_4\sigma_5 = 1 >.$$

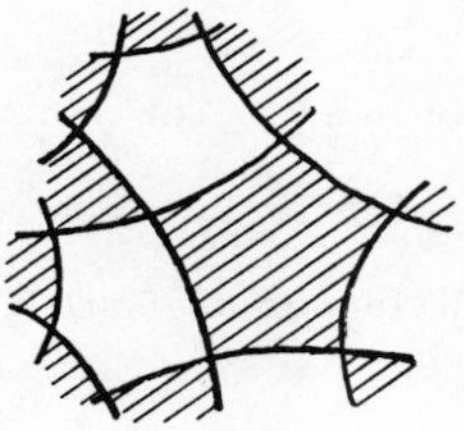

If one colors the faces Πg white for $g \in G^+$ and black for $g \notin G^+$, the result is a pentagonal checkerboard: two faces with a side in common have opposite colors, and two faces with the same color meet at most diagonally at a vertex. The union of one white and one black face Δg is a fundamental region for G^+.

Problems

Problem 1. Represent the group E_ℓ by transformations of the complex plane. How is E_ℓ related to the symmetry group Sym ℓ of the line ℓ?

Problem 2. Find the symmetry group of the tessellation of the plane exemplified by the usual pattern for brickwork, as shown.

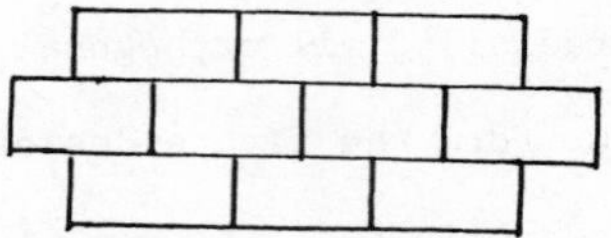

Problem 3. Show that a group G of isometries of the Euclidean plane is discontinuous if and only if any one of the following equivalent conditions is fulfilled.

(1) No point P is the limit of a sequence of distinct points Pg_n, $g_n \in G$.

(2) In the matrix representation for G, no sequence of matrices $M(g_n) \neq I$, $g_n \in G$, converges elementwise to the identity matrix I.

(3) For all $N > 0$, there are only finitely many M(g), $g \in G$, all of whose entries have absolute value less than N.

Problem 4. Let T be a group of translations of the plane, free abelian of rank 2. Show that, for all pairs α, β of generators for T, the parallelograms Π with vertices 0, 0α, 0β, $0\alpha\beta$ all have the same area.

Problem 5. Give a proof that if G is a triangle group, then G^+ has a presentation $G^+ = \langle x,y,z : x^p = y^q = z^r = xyz = 1 \rangle$.

Problem 6. What is the group G generated by reflections in the sides of a square? What is G^+? The same question for a regular hexagon.

Problem 7. The symmetry group G of a regular dodecahedron (or of a regular icosahedron) has order 120, whence G^+ has order 60. The alternating group A_5, of all even permutations of a set of five objects also has order 60. Show that $G^+ \simeq A_5$. (Reference: Coxeter-Moser, Section 4.2.)

Problem 8. Let G be the symmetry group of a regular tetrahedron. Show that G contains a subgroup H with a single face Π as fundamental region, and that $G = G_\Pi \cdot H$, but that H cannot be chosen normal in G, whence G is not a semidirect product of H by G_Π. Show that G^+ contains no subgroup H with Π as fundamental region.

Problem 9. Give a detailed proof that every closed path in the graph C can be reduced to a trivial path by successively deleting 'spines' and replacing an arc on the boundary of some Δ^* by the complementary arc.

Notes

Note 1. The icosahedral group $G^+ \simeq A_5$, for G the symmetry group of an icosahedron, is simple: it contains no normal subgroups except 1 and G^+ itself, or, alternatively, every homomorphism from G^+ onto another group H is either an isomorphism from G^+ onto H or is trivial, mapping all elements of G^+ to 1. The icosahedral group is noteworthy as the smallest nonabelian simple group.

Note 2. The commonest application of Poincaré's method is to Fuchsian groups (see Chapter 10), which contain no reflections. In this case it remains true that G is generated by all elements α that carry Δ to a region Δα having a side in common with Δ.

Note 3. A combinatorial tessellation F of E (or S or H), of type (n,m) where n, m $\geqslant$ 3, can be defined as follows.

(1) The space E (or S or H) is the nonoverlapping union of faces Δ, each homeomorphic to a closed disc.

(2) If Δ_1 and Δ_2 are two different faces, either they are disjoint, their intersection is a single point V on the boundary of both, called a vertex, or their intersection is a side s, homeomorphic to a closed interval, contained in the boundary of both, and with vertices as endpoints.

(3) The boundary of each face is a nonoverlapping union of n sides.

(4) Each vertex is contained in exactly m faces.

It is easy to see that such a tessellation must be finite just in case (n,m) is one of (3,3), (3,4), (4,3), (3,5), (5,3), and in these cases such a tessellation F of the sphere S exists and is unique within homeomorphism.

Since E and H are homeomorphic (although with different metrics), this definition does not distinguish between them. It is easy to see that for E, or H, tessellations of the remaining types exist, and are unique up to homeomorphism. We have seen that only the types (3,6), (4,4), (6,3) are realizable as regular tessellations with the Euclidean metric, and we shall see later that the remainder are realizable as regular tessellations of the hyperbolic plane H, with its metric.

References

For regular (and other) tessellations, regular (and other) solids, and their groups, see Coxeter-Moser and Coxeter's Regular Polytopes. As noted, Poincaré's method and also hyperbolic geometry will be discussed more fully in Chapters 9 and 10, where references will be given.

CHAPTER FOUR: DISCONTINUOUS GROUPS OF ISOMETRIES OF THE EUCLIDEAN PLANE: PLANE CRYSTALLOGRAPHIC GROUPS

1. Introduction

We recall that a subgroup G of the group $\mathcal{E}$ of all isometries of the Euclidean plane E is discontinuous if and only if for each point P and each circle Γ, there are only finitely many images $P\gamma$ of P, under $\gamma \in G$, inside the circle Γ. We assume henceforth that G is a discontinuous subgroup of $\mathcal{E}$. As before, $G^+ = G \cap \mathcal{E}^+$, of index 1 or 2 in G, is the normal subgroup of all elements of G that preserve orientation. Likewise, $T = G \cap \mathcal{T}$ is the normal subgroup of all translations in G.

It was seen that T is trivial, infinite cyclic, or free abelian of rank 2, and that if T is trivial G is a finite group, cyclic or dihedral, while if T is infinite cyclic G is of one of the seven types of frieze group. If T is free abelian of rank 2, then G is called a (plane) crystallographic group. We assume henceforth that G is a plane crystallographic group.

Our goal is to classify the plane crystallographic groups up to isomorphism. One might reasonably ask for a finer classification, taking into account the geometric action of G on E. For example, if G_1 and G_2 are conjugate in $\mathcal{E}$, that is, $G_2 = G_1^{\gamma}$ for some $\gamma \in \mathcal{E}$, one can think of G_2 as obtained from G_1 by a change of (rectangular) coordinates, and one would surely classify G_1 and G_2 as geometrically equivalent. But consider the case that G = T, where G is a group of translations generated by translations α and β through distances a and b in directions making an angle θ. It is easy to see that there are infinitely many of these groups not conjugate within E, whence the proposed classification is too fine to be useful. The same difficulty arises with classification according to conjugacy within the group S of similarities. However, all these translation groups $G = T$ are conjugate within the affine group A: it is easy to find γ in A carrying the vectors $\overrightarrow{0,0\alpha}$ and $\overrightarrow{0,0\beta}$, for some point 0, to two orthogonal unit vectors $\overrightarrow{e_1}$ and $\overrightarrow{e_2}$. The somewhat surprising fact is that this classification of all plane crystallographic groups by affine conjugacy coincides with the classification by isomorphism: if G_1 and G_2 are isomorphic crystallographic groups, we shall see in the sequel that there exists γ in A such that $G_2 = G_1^{\gamma}$.

Since T is a normal subgroup of G, for all γ in G the conjugation map $\tau \longmapsto \tau^{\gamma}$ is an automorphism of T, and the knowledge of this map ϕ from G into Aut T gives us almost, but not quite, complete information about G. More exactly, we have seen that $E = E_0 \cdot T$, the semidirect product of T by the stabilizer E_0 of a point 0. It happens often, <u>but not always</u>, that, correspondingly, $G = G_0 \cdot T$, semidirect product, in which case G is fully determined by the induced map G_0 into Aut T. (The exceptions arise from those glide reflections ρ in G that do not decompose <u>within G</u> into the product of a reflection $\rho_0 \in G_0$, for 0 the center of a rotation of maximal order, and a translation $\tau_0 \in T$.) This problem does not arise with G^+, which can always be written as $G^+ = G_0^+ \cdot T$, a semidirect product.

Our plan is as follows. First we classify the possible orientation preserving groups G^+. Next we use the fact that, if $G \neq G^+$, then G is generated by G^+ together with any element ρ of G that is not in G^+. We then examine the automorphism of T effected by conjugation by ρ. With this information we proceed to the detailed enumeration of cases.

2. The group G^+

We recall that a pair of generators α and β for T can be chosen such that $|\alpha| = d(0,0\alpha)$ is minimal among all $|\tau|$ for nontrivial τ in T, and such that $|\beta|$ is minimal among all $|\tau|$ for τ in T but not a power of α. We shall invoke this hypothesis on α and β whenever convenient.

When a base point 0 has been chosen, the group T determines a lattice L, the orbit $L = \{0\tau : \tau \in T\}$ of 0 under T, and indeed the set V of vectors $\overrightarrow{0,0\tau}$, $\tau \in T$, as a subset of $V(2,\underline{R})$ does not depend on the choice of 0. If γ is any element of E, then the vector $\overrightarrow{0,0\tau^\gamma}$ is the image of the vector $\overrightarrow{0,0\tau}$ under the action of γ on the plane E.

Our first result is that a rotation contained in a crystallographic group can have only one of the orders 1, 2, 3, 4, or 6. (This limitation, especially the exclusion of 5, is sometimes called the 'crystallographic restriction'.)

THEOREM. G^+ <u>is generated by</u> T <u>together with a single rotation of order</u> n, <u>for</u> n = 1, 2, 3, 4, <u>or</u> 6, <u>and is of one of the following types</u>:

$G_1 = \langle \alpha,\beta : \alpha\beta = \beta\alpha \rangle$;

$G_2 = \langle \alpha,\beta,\sigma : \alpha\beta = \beta\alpha,\ \sigma^2 = 1,\ \alpha^\sigma = \alpha^{-1},\ \beta^\sigma = \beta^{-1} \rangle$;

$G_3 = \langle \alpha,\beta,\sigma : \alpha\beta = \beta\alpha,\ \sigma^3 = 1,\ \alpha^\sigma = \alpha^{-1}\beta,\ \beta^\sigma = \alpha^{-1} \rangle$;

$G_4 = \langle \alpha,\beta,\sigma : \alpha\beta = \beta\alpha,\ \sigma^4 = 1,\ \alpha^\sigma = \beta,\ \beta^\sigma = \alpha^{-1} \rangle$;

$G_6 = \langle \alpha,\beta,\sigma : \alpha\beta = \beta\alpha,\ \sigma^6 = 1,\ \alpha^\sigma = \beta,\ \beta^\sigma = \alpha^{-1}\beta \rangle$.

<u>Proof</u>. Suppose that G^+ contains a rotation σ of order $n \geqslant 2$, say with center 0. Then the n points 0α, $0\alpha\sigma,\dots,0\alpha\sigma^{n-1}$ are evenly spaced around the circle at 0 of radius $|\alpha|$. If $n > 6$, then $d(0\alpha,0\alpha\sigma) < |\alpha|$. Since α is in T, α^σ and $\alpha^{-1}\alpha^\sigma$ are in T. But $\alpha^{-1}\alpha^\sigma$ carries 0α to $0\alpha^\sigma = 0\sigma^{-1}\alpha\sigma = 0\alpha\sigma$, hence has length

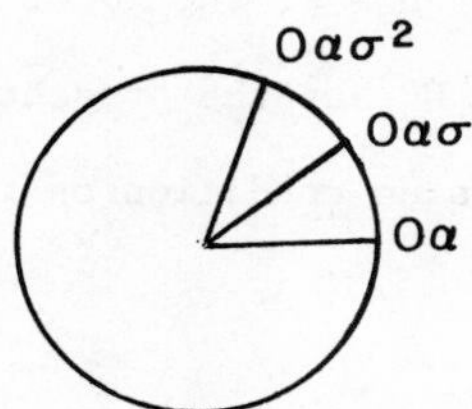

$|\alpha^{-1}\alpha^{\sigma}| < |\alpha|$, contrary to the choice of α.

If $n = 5$, $0\alpha^{-1}\sigma^3$ is one tenth of the way around the circle from 0α, whence $d(0\alpha, 0\alpha^{-1}\sigma^3) < |\alpha|$, and the translation $\alpha^{-1}(\alpha^{-1})^{\sigma^3}$ has length less than $|\alpha|$, again a contradiction. Thus all rotations in G^+ have order 1, 2, 3, 4, or 6.

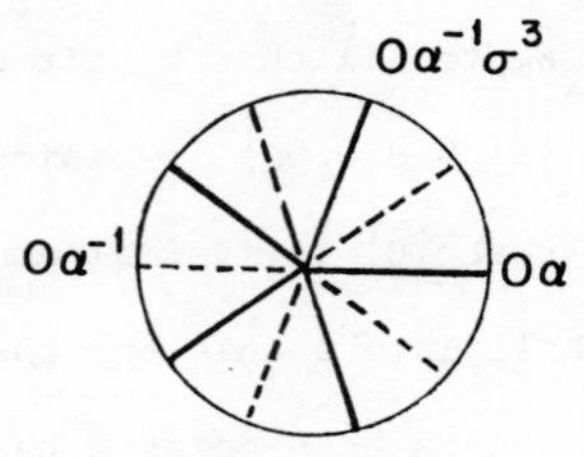

If G^+ contains a rotation σ_1 with angle $2\pi/2$ and a rotation σ_2 with angle $2\pi/3$, then $\sigma_1\sigma_2^{-1}$, is a rotation with angle $2\pi/6$. Thus if G^+ contains rotations of orders 2 and 3 it contains rotations of orders 6. If G^+ contained σ_1 with angle $2\pi/3$ and σ_2 with angle $2\pi/4$, then $\sigma_1\sigma_2^{-1}$ would be a rotation with angle $2\pi/12$, of order 12. Thus G^+ cannot contain rotations of orders both 3 and 4. We conclude that if n is the greatest order of a rotation σ in G^+, then all other rotations σ_1 in G^+ have orders n_1 dividing n. We may suppose that σ has angle $2\pi/n$ and σ_1 has angle $2\pi/n_1$, where $n = n_1 m$. Then σ^m has the same angle $2\pi/n_1$ as σ_1, and $\sigma_1\pi^{-m} = \tau$ is a translation. Thus σ_1 is in the group generated by T together with σ. This shows that G^+ is generated by T together with σ.

We continue to let $n = 1, 2, 3, 4$, or 6 be the greatest order of a rotation in G^+, and write G_n for G^+ according to the case.

If $n = 1$, then $G^+ = G_1$ contains no nontrivial rotation, and $G_1 = T$, with the presentation given.

If $n = 2$, then σ of order 2 has angle π, whence $\tau^{\sigma} = \tau^{-1}$ for all τ in T, and G_2 has the presentation given. The presence of σ of order 2 imposes no condition on the lattice L.

Let $n = 4$, with σ a rotation of order 4 and center 0. Since σ has angle $\pi/2$, α^σ is a translation in a direction perpendicular to that of α, with length $|\alpha^\sigma| = |\alpha|$. We may choose $\beta = \alpha^\sigma$. Then $\beta^\sigma = \alpha^{-1}$, and G_4 has the given presentation. In this case the lattice L is a <u>square lattice</u>.

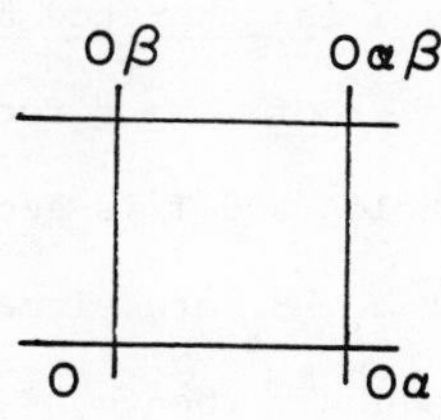

Suppose now that G^+ contains a rotation σ of order 3, with center 0. Since the angle from $\overrightarrow{0,0\alpha}$ to $\overrightarrow{0,0\alpha^\sigma}$ is $2\pi/3$, and $|\alpha^\sigma| = |\alpha|$, the three points 0, $0\alpha^\sigma$, $0\alpha^\sigma\alpha$ are the vertices of an equilateral triangle. Thus $|\alpha^\sigma\alpha| = |\alpha|$ and we may choose $\beta = \alpha^\sigma\alpha$, whence $\alpha^\sigma = \beta\alpha^{-1}$ and $\beta^\sigma = \alpha^{-1}$. In this case the lattice L is a <u>triangular lattice</u>.

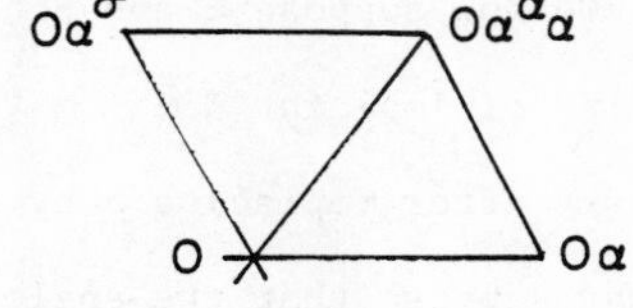

If $n = 3$, then G_3, generated by T together with σ of order 3, has the given presentation.

If $n = 6$, then σ^2 has order 3, whence the lattice L is triangular. However σ now has angle $2\pi/6$, whence $\alpha^\sigma = \beta$ and $\beta^\sigma = \beta\alpha^{-1}$, in accordance with the given presentation. □

3. Conjugation of T by ρ

We now turn to the case that $G \neq G^+$, whence G is generated by G^+ together by any element ρ of G that is not in G^+. Now ρ is a reflection or a glide reflection, and in either case ρ^2 is in T. Thus the automorphism $\tau \longmapsto \tau^\rho$ of T, which cannot fix both α and β, is of order 2.

LEMMA. <u>If</u> $\rho \in G$, $\rho \notin G^+$, <u>then one or the other, but not both, of the following conditions holds:</u>

(A1) T <u>is generated by elements</u> α <u>and</u> β <u>such that</u> $\alpha^\rho = \alpha$ <u>and</u> $\beta^\rho = \beta^{-1}$;

(A2) T <u>is generated by elements</u> α <u>and</u> β <u>such that</u> $\alpha^\rho = \beta$ <u>and</u> $\beta^\rho = \alpha$.

<u>Proof</u>. We first show that A1 and A2 are incompatible. Suppose that A2 holds and T is generated by α and β such that $\alpha^\rho = \beta$ and $\beta^\rho = \alpha$, and that α_1, β_1 are elements of T such that $\alpha_1^\rho = \alpha_1$ and $\beta_1^\rho = \beta_1^{-1}$. If $\alpha_1 = \alpha^h\beta^k$, then $\alpha_1^\rho = \alpha^k\beta^h$, and $\alpha_1^\rho = \alpha_1$ implies that $h = k$ and $\alpha_1 = (\alpha\beta)^h$. If $\beta_1 = \alpha^m\beta^n$, then $\beta_1^\rho = \alpha^n\beta^m$, and $\beta_1^\rho = \beta_1^{-1}$ implies that $m = -n$ and $\beta_1 = (\alpha\beta^{-1})^m$. Now α_1 and β_1 are both in the subgroup generated by $\alpha\beta$ and $\alpha\beta^{-1}$, which is a proper subgroup of T. Thus α_1 and β_1 do not generate T.

We now suppose α and β chosen as before, with $|\alpha|$ minimal for $\alpha \neq 1$ and $|\beta|$ minimal for $\beta \notin \langle \alpha \rangle$. We now impose some supplementary conditions. After replacing α by α^{-1} if necessary, we may suppose that the angle from $\overrightarrow{0,0\alpha}$ to $\overrightarrow{0,0\beta}$ is some $\theta \leq \frac{\pi}{2}$. The minimality of $|\beta|$ now requires that $d(0,0\beta) \leq d(0\alpha,0\beta)$, whence the component of $\overrightarrow{0,0\beta}$ parallel to $\overrightarrow{0,0\alpha}$ has length $h(\beta) \leq \frac{1}{2}|\alpha|$, that is, 0β is either on the perpendicular bisector of $[0,0\alpha]$ or on the same side of it as 0.

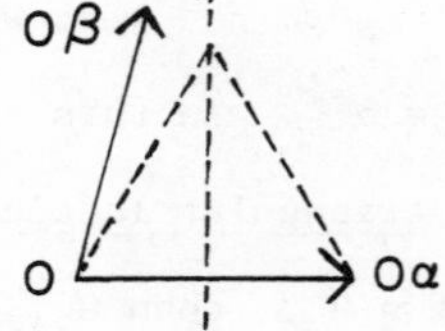

This, together with the condition that $|\beta| \geq |\alpha|$, implies that $\theta \geq \frac{\pi}{3}$. Thus $\frac{\pi}{3} \leq \theta \leq \frac{\pi}{2}$.

Suppose first that $\alpha^\rho = \alpha$, whence ρ has its axis parallel to α. If $\theta = \frac{\pi}{2}$, then $\beta^\rho = \beta^{-1}$ and A1 holds. If $\theta < \frac{\pi}{2}$, then $\overrightarrow{0,0\beta^\rho}$ has the same component perpendicular to $\overrightarrow{0,0\alpha}$ as $\overrightarrow{0,0\alpha\beta^{-1}}$, while it has the same component, $h(\beta^\rho) = h(\beta)$, as $\overrightarrow{0,0\beta}$ parallel to $\overrightarrow{0,0\alpha}$. Since $0 < h(\beta) \leq \frac{1}{2}|\alpha|$ implies that $\frac{1}{2}|\alpha| \leq h(\alpha\beta^{-1}) < |\alpha|$, we conclude that β^ρ can only be $\beta^\rho = \alpha\beta^{-1}$. Let

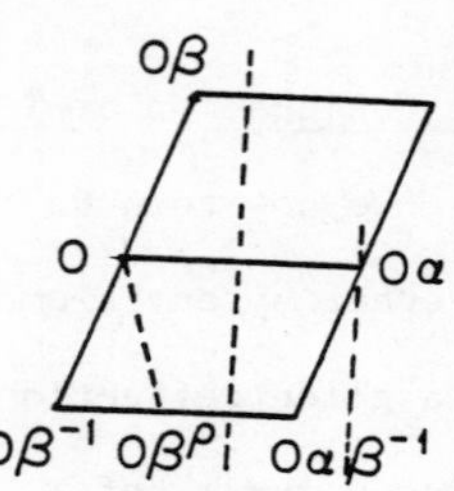

$\alpha_1 = \beta$ and $\beta_1 = \alpha\beta^{-1}$; then α_1 and β_1 generate T while ρ exchanges them, and A2 holds.

Suppose next that $\alpha^\rho = \alpha^{-1}$, whence ρ has its axis perpendicular to α. If $\theta = \frac{\pi}{2}$, then $\beta^\rho = \beta$, and A1 holds, with α and β exchanged. If $\theta < \frac{\pi}{2}$, then $\overrightarrow{0\beta^\rho, 0\beta}$ is parallel to $\overrightarrow{0, 0\alpha}$ and of length at most $2h(\beta) \leqslant |\alpha|$, whence, by minimality of $|\alpha|$, it must have length $|\alpha|$, and $\beta^\rho = \beta\alpha^{-1}$.

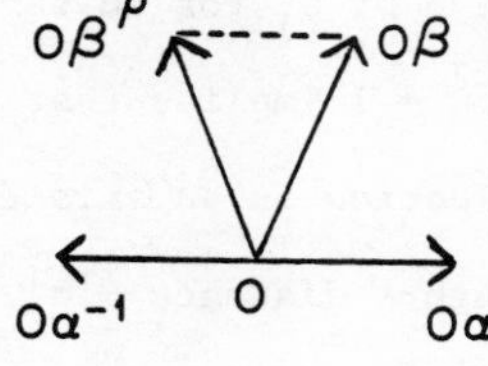

Now ρ exchanges the pair of generators β and $\beta\alpha^{-1}$ of T, and A2 holds.

Suppose finally that $\alpha^\rho \neq \alpha, \alpha^{-1}$, whence α^ρ, of length $|\alpha^\rho| = |\alpha|$, is not a power of α. We take $\beta = \alpha^\rho$. Even though the supplementary condition need not be fulfilled, the minimality conditions are, whence α and β generate T and A2 holds. □

4. Enumeration of cases

We now enumerate the possible types of $G \neq G^+ = G_n$, according to $n = 1, 2, 3, 4, 6$.

Case n = 1. Here $G_1 = T$. If ρ_1, ρ_2 are elements of G not in G^+, then $\rho_2 = \rho_1\tau$ for some τ in T, whence $\tau^{\rho_1} = \tau^{\rho_2}$ for all τ in T. In short, all choices of ρ yield the same automorphism of T.

If G contains any reflection, we choose ρ to be a reflection, with $\rho^2 = 1$. It is clear that both cases A1 and A2 can be realized, for example (after an affine transformation), with the square lattice L. We obtain two groups G_1^1 and G_1^2 with presentations obtained from that for G_1 by adjoining the new generator ρ, a relation $\rho^2 = 1$, and two relations giving the values of α^ρ and β^ρ.

Suppose now that G contains no reflection, whence we must choose ρ as a glide reflection with $\rho^2 = \tau \neq 1$ in T. If A1 holds, then $\tau^\rho = \tau$ implies that $\tau = \alpha^h$ for some integer h. If $\rho_1 = \rho\alpha^m$, then $\rho_1^2 = \alpha^{h+2m}$, and, replacing ρ by ρ_1 for suitable m we may suppose that $\rho^2 = 1$ or $\rho^2 = \alpha$. Since $\rho^2 = 1$ implies that ρ is a reflection, we must have $\rho^2 = \alpha$. Thus ρ is reflection in an axis ℓ parallel to α followed by a translation along ℓ through a distance $\frac{1}{2}|\alpha|$. We must verify that this group G_1^3 is new, that is, that it contains no reflection. Now every element of $G = G_1^3$ not in G^+ has the form $\rho_2 = \rho\alpha^m\beta^n$ for some integers m and n, and we find that $\rho_2^2 = \rho\alpha^m\beta^n\rho\alpha^m\beta^n = \rho^2\alpha^m\beta^{-n}\alpha^m\beta^n = \alpha^{2m+1} \neq 1$, whence ρ_2 is not a reflection.

There remains the possibility that G contains no reflection and A2 holds. Now $\rho^2 = \tau$ and $\tau^\rho = \tau$ implies that $\tau = (\alpha\beta)^h$ for some h, and, as above, we can suppose that $\rho^2 = \alpha\beta$. Now let $\rho_2 = \rho\alpha^{-1}$. Then $\rho_2^2 = \rho\alpha^{-1}\rho\alpha^{-1} = \rho^2(\alpha^\rho)^{-1}\alpha^{-1} = (\alpha\beta)\beta^{-1}\alpha^{-1} = 1$, and G contains the reflection ρ_2, contrary to hypothesis.

THEOREM. _If_ $G \neq G^+ = G_1$, _there are exactly three possible isomorphism types for_ G, _as follows_:

$$G_1^1 = \langle \alpha,\beta,\rho : \alpha\beta = \beta\alpha,\ \alpha^\rho = \alpha,\ \beta^\rho = \beta^{-1},\ \rho^2 = 1 \rangle;$$
$$G_1^2 = \langle \alpha,\beta,\rho : \alpha\beta = \beta\alpha,\ \alpha^\rho = \beta,\ \beta^\rho = \alpha,\ \rho^2 = 1 \rangle;$$
$$G_1^3 = \langle \alpha,\beta,\rho : \alpha\beta = \beta\alpha,\ \alpha^\rho = \alpha,\ \beta^\rho = \beta^{-1},\ \rho^2 = \alpha \rangle.$$

Case n = 2. Here G_2 is generated by α, β, σ where $\sigma^2 = 1$ and $\alpha^\sigma = \alpha^{-1}$, $\beta^\sigma = \beta^{-1}$.

Suppose first that G contains a reflection ρ and that A1 holds, with $\alpha^\rho = \alpha$, $\beta^\rho = \beta^{-1}$. This implies that α and β are perpendicular. Then ρ has its axis parallel to α. Let $\rho_1 = \sigma\rho$. Since $\beta^{\rho_1} = \beta$, ρ_1 has its axis parallel to β, with $\beta_1^2 = \beta^h$ for some integer h. Replacing ρ by $\rho\beta^m$ for some integer m, we can suppose, according to the parity of h, that $\rho_1^2 = 1$

or $\rho_1^2 = \beta$. If ρ and ρ' are any two reflections parallel to α, then $\rho'\rho = \beta^m$ for some m, and $\rho' = \rho\beta^m$, whence different choices of ρ yield the same case $\rho_1^2 = 1$ or $\rho_1^2 = \beta$, and the two cases are distinct. These two cases G_2^1 and G_2^2 are easily realized with the square lattice L. For the first we take ρ with axis $\ell = \overline{0,0\alpha}$.

For the second, we take ℓ parallel to $\overline{0,0\alpha}$, at a distance of $\frac{1}{4}|\beta|$ in the direction of 0β. Then

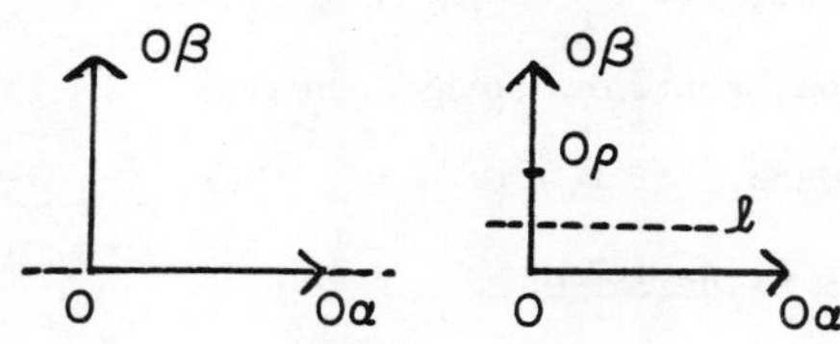

$0\rho_1 = 0\sigma\rho = 0\rho$ is the midpoint of $[0,0\beta]$, and ρ_1, with axis $\overline{0,0\beta}$ carries 0 to 0ρ, whence $\rho_1^2 = \beta$.

Suppose now that G contains a reflection ρ and that A2 holds, with $\alpha^\rho = \beta$, $\beta^\rho = \alpha$, whence ρ has axis parallel to $\alpha\beta$. If $\rho_1 = \sigma\rho$, then $\alpha^{\rho_1} = \beta^{-1}$, $\beta^{\rho_1} = \alpha^{-1}$, and $(\alpha^{-1}\beta)^{\rho_1} = \alpha^{-1}\beta$, whence ρ_1 has axis parallel to $\alpha^{-1}\beta$, and $\rho_1^2 = (\alpha^{-1}\beta)^h$ for some integer h. Replacing ρ by $\rho(\alpha^{-1}\beta)^m$ for some m, we may suppose, according to the parity of h, that $\rho_1^2 = 1$ or $\rho_1^2 = \alpha^{-1}\beta$. If $\rho_1^2 = \alpha^{-1}\beta$, let $\rho_2 = \rho_1\alpha$. Then $\rho_2^2 = \rho_1\alpha\rho_1\alpha = \rho_1^2\beta^{-1}\alpha = (\alpha^{-1}\beta)\beta^{-1}\alpha = 1$, and ρ_2 is a reflection. In either case G contains two reflections with perpendicular axes parallel to $\alpha\beta$ and $\alpha^{-1}\beta$ (the lines containing the diagonal of the rhombus Π) whose product is a rotation of order 2, and the group $G_2^3 = G_0 \cdot T$, semidirect product, with $G_0 \simeq D_4$.

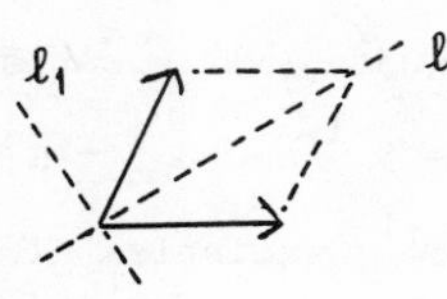

Finally, suppose that G contains no reflection. If A1 holds, we may suppose that $\rho^2 = \alpha$. If $\rho_1 = \sigma\rho$, with $\sigma^{\rho_1} = \alpha^{-1}$, $\beta^{\rho_1} = \beta$, we may likewise suppose that $\rho_1^2 = \beta$. To check that the group G_2^4 defined thus contains no reflections, we verify that $(\rho\alpha^m\beta^n)^2 = \alpha^{2m+1} \neq 1$

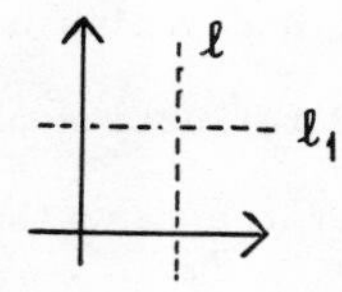

and that $(\rho_1\alpha^m\beta^n)^2 = \beta^{2n+1} \neq 1$. This group is easily realized in the square lattice L by taking ρ and ρ_1 with axes parallel to $\overline{0,0\alpha}$ and $\overline{0,0\beta}$, at a distance $\frac{1}{2}|\alpha| = \frac{1}{2}|\beta|$, and with translational displacement $\frac{1}{2}|\alpha| = \frac{1}{2}|\beta|$.

If A2 holds, with $\alpha^\rho = \beta$, $\beta^\rho = \alpha$, we may suppose, as before, that $\rho^2 = \alpha\beta$ and, for $\rho_1 = \sigma\rho$, that $\rho_1^2 = \alpha^{-1}\beta$. As before, $\rho_2 = \rho_1\alpha$ is then a reflection, contrary to hypothesis.

THEOREM. <u>If</u> $G \neq G^+ = G_2$, <u>there are exactly four possible isomorphism types for</u> G, <u>as follows</u>:

$$G_2^1 = \langle\, \alpha,\beta,\rho \,:\, \alpha\beta = \beta\alpha,\ \alpha^\sigma = \beta,\ \beta^\sigma = \alpha^{-1},\ \alpha^\rho = \alpha,\ \beta^\rho = \beta^{-1},\ \rho^2 = 1,\ (\sigma\rho)^2 = 1 \,\rangle;$$

$$G_2^2 = \langle\, \alpha,\beta,\rho \,:\, \alpha\beta = \beta\alpha,\ \alpha^\sigma = \beta,\ \beta^\sigma = \alpha^{-1},\ \alpha^\rho = \alpha,\ \beta^\rho = \beta^{-1},\ \rho^2 = 1,\ (\sigma\rho)^2 = \beta \,\rangle;$$

$$G_2^3 = \langle\, \alpha,\beta,\rho \,:\, \alpha\beta = \beta\alpha,\ \alpha^\sigma = \beta,\ \beta^\sigma = \alpha^{-1},\ \alpha^\rho = \beta,\ \beta^\rho = \alpha,\ \rho^2 = 1,\ (\sigma\rho)^2 = 1 \,\rangle;$$

$$G_2^4 = \langle\, \alpha,\beta,\rho \,:\, \alpha\beta = \beta\alpha,\ \alpha^\sigma = \beta,\ \beta^\sigma = \alpha^{-1},\ \alpha^\rho = \alpha,\ \beta^\rho = \beta^{-1},\ \rho^2 = \alpha,\ (\sigma\rho)^2 = \beta \,\rangle.$$

<u>Case n = 4</u>. We have $\sigma^4 = 1$ with $\alpha^\sigma = \beta$, $\beta^\sigma = \alpha^{-1}$, and the lattice L is square. If ρ satisfies A1, with $\alpha^\rho = \alpha$, $\beta^\rho = \beta^{-1}$, and $\rho_1 = \sigma\rho$, then $\alpha^{\rho_1} = \beta^{-1}$ and $\beta^{\rho_1} = \alpha^{-1}$, whence ρ_1, which exchanges the two generators α and β^{-1}, satisfies A2. Thus the cases A1 and A2 coincide, and we may suppose that ρ satisfies A1, where ρ is either a reflection or a glide reflection.

The set of four elements $\{\alpha,\alpha^{-1},\beta,\beta^{-1}\}$ is uniquely determined by the facts that σ permutes them cyclically and that they generate T. We show that if G contains a reflection parallel to one of α, β then it also contains a reflection parallel to the other. By symmetry, it suffices to consider the case that G contains a reflection ρ parallel to α, hence satisfying A1. Since $(\alpha^{-1}\beta)^{\rho_1} = \alpha^{-1}\beta$, the axis of ρ_1 is parallel to $\alpha^{-1}\beta$, and $\rho_1^2 = (\alpha^{-1}\beta)^h$ for some integer h. Now $\rho_1\beta^{-h}$ is parallel to ρ_1, hence to $\alpha^{-1}\beta$, and $(\rho_1\beta^{-h})^2 = \rho_1^2\alpha^h\beta^{-h} = (\alpha^{-1}\beta)^{h-h} = 1$. We have thus two reflections with axes meeting at some point 0_1 with an angle of $\frac{2\pi}{8}$, whence their product σ_1 is a rotation about 0_1 of order 4. The group G then has the form $G_4^1 = G_{0_1}\cdot T$ where $G_{0_1} \simeq D_8$. This group is easily realized on the square lattice. In particular, it contains a reflection parallel to β.

Suppose now that G contains a glide reflection ρ parallel to α, but no reflection parallel to α. Then $G = G_4^2$ is not isomorphic to G_4^1. To find a presentation for G_4^1, after replacing ρ by some $\rho\alpha^m$ we can suppose, as before, that $\rho^2 = \alpha$. If we now replace ρ by some $\rho\beta^n$, the relation $\rho^2 = \alpha$ remains valid while, as before, for proper choice of n we can make $\rho_1^2 = 1$. Thus we have the relation $(\sigma\rho)^2 = 1$, or $\sigma^\rho = \sigma^{-1}$.

This group can be realized on the square lattice as follows. Let σ have center $0 = (0,0)$, and let $0\alpha = (1,0)$, $0\beta = (0,1)$. If ρ is chosen with axis ℓ through the point $\left(\frac{1}{4}, \frac{1}{4}\right)$, then the axis ℓ_1 of ρ_1 will also pass through this point. In the figure, points marked ● are centers of rotations of order 4, points marked **x** are centers of rotations of order 2, oblique solid lines / , \ are axes of reflections, and horizontal or vertical broken lines -----, ¦ are axes of glide reflections.

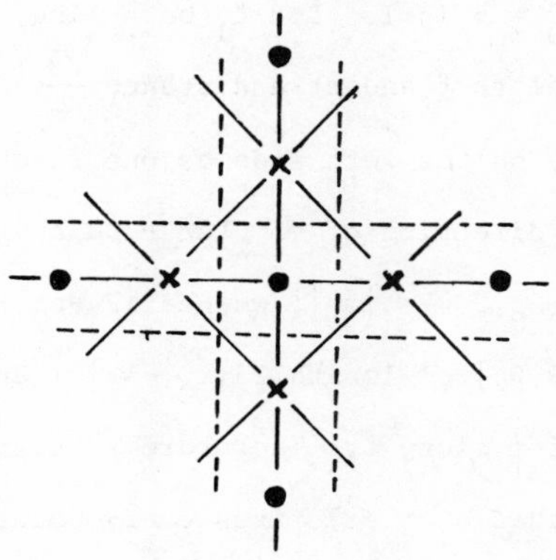

THEOREM. <u>If</u> $G \neq G^+ = G_4$, <u>there are exactly two possible isomorphism types for</u> G, <u>as follows</u>:

$$G_4^1 = \langle \alpha,\beta,\sigma,\rho : \alpha\beta = \beta\alpha,\ \sigma^4 = 1,\ \alpha^\sigma = \beta,\ \beta^\sigma = \alpha^{-1},\ \alpha^\rho = \alpha,\ \beta^\rho = \beta^{-1},$$
$$\rho^2 = 1,\ (\sigma\rho)^2 = 1 \rangle;$$

$$G_4^2 = \langle \alpha,\beta,\sigma,\rho : \alpha\beta = \beta\alpha,\ \sigma^4 = 1,\ \alpha^\sigma = \beta,\ \beta^\sigma = \alpha^{-1},\ \alpha^\rho = \alpha,\ \beta^\rho = \beta^{-1},$$
$$\rho^2 = \alpha,\ (\sigma\rho)^2 = 1 \rangle.$$

Case n = 3. Here G contains a rotation σ of order 3, and the lattice L is triangular with $\alpha^\sigma = \alpha^{-1}\beta$ and $\beta^\sigma = \alpha^{-1}$.

We show that if G contains a glide reflection ρ with axis ℓ, then G also contains a reflection ρ_1 with axis ℓ_1 parallel to ℓ. Let $\rho = \rho_0\tau_0$, where ρ_0 is reflection in ℓ and τ_0 is a translation along ℓ; here ρ_0 and τ_0 in E are not necessarily in G. Thus $\rho^2 = \tau_0^2 = \tau$ in T. Let ℓ_1 be a line parallel to ℓ and at a distance $\frac{\sqrt{3}}{2}$ from ℓ, on the left side as one faces

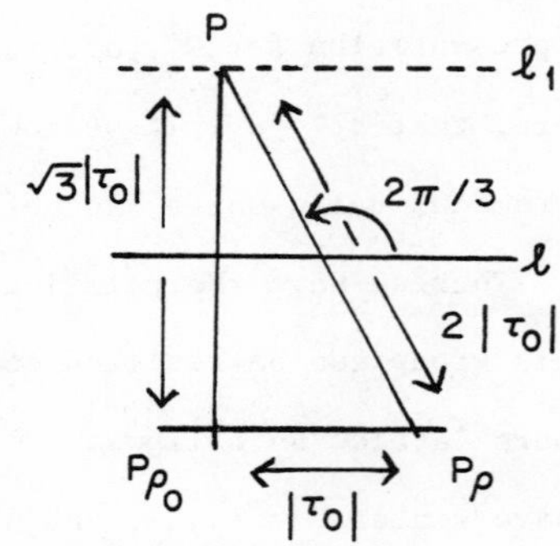

in the direction of τ. Let P be any point on ℓ_1. Then $d(P,P\rho_0) = \sqrt{3}\cdot|\tau_0|$ and $d(P\rho_0,P\rho) = |\tau_0|$, whence $[P,P\rho_0,P\rho]$ is a right triangle with hypotenuse $[P,P\rho]$ of length $2\cdot|\tau_0| = |\tau|$ and at an angle of $\frac{2\pi}{3}$ from the direction of τ along ℓ. Therefore τ^σ carries $P\rho$ to P, and $P\rho\tau^\sigma = P$. We have shown that $\rho_1 = \rho\tau^\sigma$ fixes every point P of ℓ_1, whence ρ_1 is a reflection with axis ℓ_1 parallel to ℓ.

If G is generated by $G^+ = G_3$ together with some ρ that satisfies condition A1, then the axis of ρ is parallel to α and the axis of $\sigma\rho$ is parallel to β. The result above shows that G then contains reflections ρ and ρ' with axes parallel to α and β. If the axes of ρ and ρ' meet at 0_1, then $\sigma_1 = \rho\rho'$ is a rotation about 0_1 of order 3. Thus $G = G_{0_1}\cdot T$, semidirect product, where $G_{0_1} \simeq D_6$.

If ρ satisfies A2, then it has axis parallel to $\alpha\beta$, and the same argument shows again that $G = G_{0_1}\cdot T$, semidirect product, but now with the three axes of reflection through 0_1 in a different position relative to $\overrightarrow{0,0\alpha}$, $\overrightarrow{0,0\beta}$. We have thus two types G_3^1 and G_3^2.

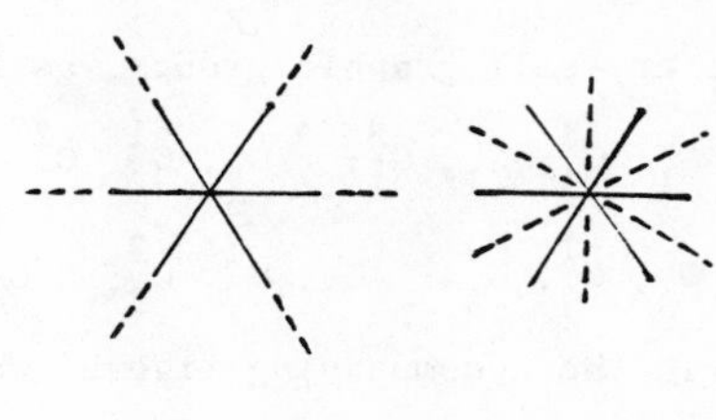

THEOREM. <u>If</u> $G \neq G^+ = G_3$, <u>there are exactly two possible isomorphism types for</u> G, <u>as follows</u>:

$$G_3^1 = \langle\, \alpha,\beta,\sigma,\rho \;:\; \alpha\beta = \beta\alpha,\ \sigma^3 = 1,\ \alpha^\sigma = \alpha^{-1}\beta,\ \beta^\sigma = \alpha^{-1},$$
$$\alpha^\rho = \alpha,\ \beta^\rho = \alpha\beta^{-1},\ \rho^2 = 1,\ (\sigma\rho)^2 = 1 \,\rangle;$$
$$G_3^2 = \langle\, \alpha,\beta,\sigma,\rho \;:\; \alpha\beta = \beta\alpha,\ \sigma^3 = 1,\ \alpha^\sigma = \alpha^{-1}\beta,\ \beta^\sigma = \alpha^{-1},$$
$$\alpha^\rho = \beta,\ \beta^\sigma = \alpha,\ \rho^2 = 1,\ (\sigma\rho)^2 = 1 \,\rangle.$$

<u>Case n = 6</u>. If ρ is a reflection or glide reflection with axis ℓ, then $\rho_1 = \sigma\rho$ has an axis ℓ_1 making an angle of $\frac{2\pi}{12}$ with ℓ. Since G contains the rotation σ^2 of order 3, from the discussion of the case $n = 3$ we conclude that G contains reflections ρ and ρ' with axes making an angle of $\frac{2\pi}{12}$, whence $\sigma_1 = \rho\rho'$ is a rotation of order 6 about their point of intersection 0_1. Thus G has the form $G_6^1 = G_{0_1}\cdot T$, semidirect product, where $G_{0_1} \simeq D_{12}$. Among the six reflections with axes passing through 0_1, there is one satisfying A1 and another satisfying A2, whence G is always of the form G_6^1.

THEOREM. <u>If</u> $G \neq G^+ = G_6$, <u>then there is exactly one isomorphism type for</u> G, <u>as follows</u>:

$$G_6^1 = \langle\, \alpha,\beta,\sigma,\rho \;:\; \alpha\beta = \beta\alpha,\ \sigma^6 = 1,\ \alpha^\sigma = \beta,\ \beta^\sigma = \alpha^{-1}\beta,$$
$$\alpha^\rho = \alpha,\ \beta^\rho = \alpha\beta^{-1},\ \rho^2 = 1,\ (\sigma\rho)^2 = 1 \,\rangle.$$

5. Summary

We have seen that there are in all seventeen isomorphism types of plane crystallographic groups, as follows:

G_1, G_1^1, G_1^2, G_1^3; G_2, G_2^1, G_2^2, G_2^3, G_2^4;

G_4, G_4^1, G_4^2; G_3, G_3^1, G_3^2; G_6, G_6^1.

In the accompanying figure, we display seventeen 'patterns' in the plane whose symmetry groups exemplify these seventeen types. By a pattern we mean a figure F in the plane E. First of all, F consists of lines dividing the plane into squares or equilateral triangles Q, with vertices the points of the square or triangular lattice L. In addition, F consists of certain sets 'inscribed' in the cells Q, such as a conventionalized letter of the alphabet or a circle, to reduce the symmetry group of F to a suitable subgroup of the full symmetry group of the tessellation of E by the cells Q. In the figure we have shown of necessity only a small part of the pattern F, which is to be extended periodically in the obvious way.

6. Notes and references

Crystallographic groups of dimensions 2 and 3 first arose from their obvious connections with chemistry and physics. They came into mathematical prominence in 1900 with D. Hilbert's famous list of important outstanding mathematical problems. The main part of Hilbert's 18th problem was solved by L. Bieberbach in 1910, who showed that the number of (analogously defined) types of crystallographic groups in Euclidean space of any dimension $n \geqslant 1$ is finite. In fact, for dimensions 1, 2, 3, and 4 the number is 2, 17, 219, and 4783.

For an accessible exposition of the theory of n-dimensional crystallographic groups, see R. L. E. Schwartzenberger (1). For a brief commen-

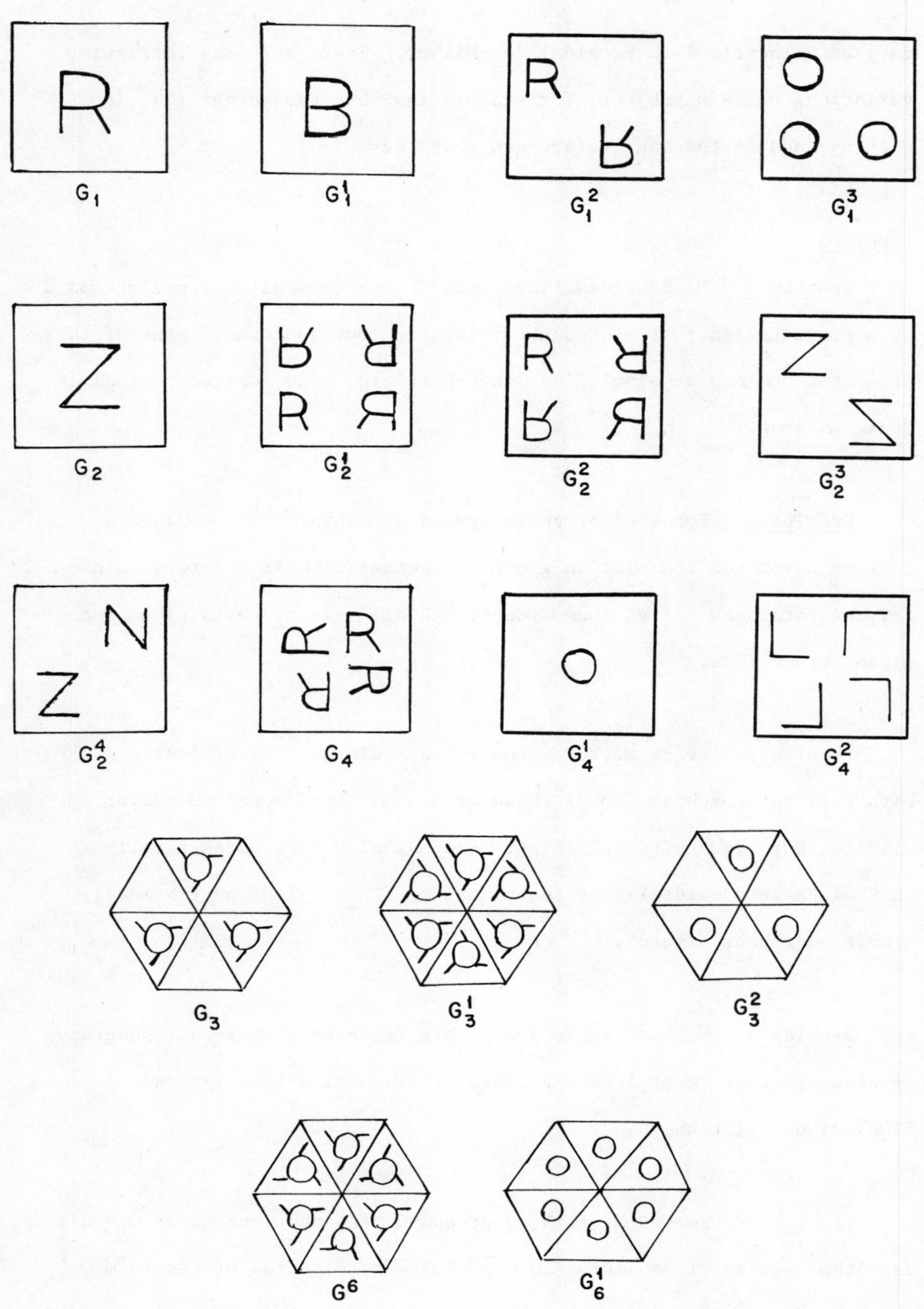
G1
G1¹
G1²
G1³
G2
G2¹
G2²
G2³
G2⁴
G4
G4¹
G4²
G3
G3¹
G3²
G6
G6¹

tary on Hilbert's 18th problem, see Milnor. There are many intriguing variations on this subject, for example in Schwartzenberger (2), in Wieting, and in the many references that they cite.

Problems

Problem 1. We have described each of the crystallographic groups G by a presentation that arises naturally from our approach. Many of these presentations can be simplified; when possible, find more economical presentations.

Problem 2. For each of these groups G, exhibit in the plane a lattice L for the translation group T together with all centers of non-trivial rotations (of various orders) and all axes of reflections and of glide reflections.

Problem 3. Every author seems to have his own set of 'patterns' F with full automorphism groups these groups G. Ours were chosen for clarity, but are a bit artificial. Find simpler more geometrically natural patterns, ideally of tessellations of the plane by reasonably simple congruent figures.

Problem 4. Many of these groups are isomorphic to proper subgroups of others, or of themselves. Exhibit diagrammatically all these 'inclusion' relations.

Problem 5. For each of these groups G, the quotient group $\bar{G} = G/T$ is either cyclic of the form $< \bar{\sigma} : \bar{\sigma}^n = 1 >$ or dihedral of the form $< \bar{\sigma},\bar{\rho} : \bar{\sigma}^n = (\bar{\rho}\bar{\sigma})^2 = \bar{\rho}^2 = 1 >$, where $\bar{\sigma}$ is the image of a rotation $\bar{\sigma}$ in G and $\bar{\rho}$ of a reflection or glide reflection ρ in G. Thus $(\rho\sigma)^2$ and ρ^2 are in T. The group G, as abstract group, is determined by a knowledge of $\bar{G}$ as abstract group, of the conjugation map ϕ from $\bar{G}$ into Aut T, where T is regarded as an abstract group, together with the knowledge of the elements $(\rho\sigma)^2$ and ρ^2 in T. Use this approach to obtain a classification of the (isomorphism types) of the groups G as abstract groups.

Problem 6. Each of these groups extends to a group G^* by replacing T by the larger group $\mathcal{T}$ of all translations of the plane. Classify these (nondiscontinuous) groups G^*.

Problem 7. Which of these groups G contains glide reflections but no reflections? Which contain glide reflections which do not factor, within G, into the product of a reflection and a translation?

CHAPTER FIVE: REGULAR TESSELLATIONS IN HIGHER DIMENSIONS

1. Introduction

We have seen that there are exactly five types of regular tessellations of the 2-sphere S^2, which correspond to exactly five types of regular solids in three dimensions. We have seen also that there are exactly three types of regular tessellations of the Euclidean plane E^2.

It is natural to ask how many types of regular tessellations there are of S^n and of E^n for $n \geqslant 3$. This question has a simple answer. There are six types of regular tessellation of S^3, and only three types of regular tessellation of S^n for all $n \geqslant 4$, corresponding to the three (n+1)-dimensional regular solids that are the natural analogs of the 3-dimensional tetrahedron, the cube, and the dual to the cube, that is, the octahedron. There is only one type of regular tessellation of E^n for all $n \geqslant 3$, the natural analog of the tessellation of E^3 by 3-cubes except for $n = 4$, where there are two more.

In the interest of intuitive visualization, we have for the most part confined these notes to 2-dimensional geometry, but it turns out that to discuss the regular tessellations in higher dimensions we do not need to depart far from this limitation. The only facts from higher dimensional geometry that we use follow in the most obvious way from the representation of E^n as the 'coordinate space' of all points $P = (x_1,\ldots,x_n)$ where $x_1,\ldots,x_n$ are real numbers.

A second limitation that we have imposed on this book is to use methods of analytic geometry as sparingly as possible, avoiding, for example, any use of plane or spherical trigonometry. This limitation presents some more serious difficulties, although only in the discussion of dimensions 3 and 4. There are a few results in these dimensions that we do not know how to prove by the methods to which we have restricted ourselves. These results we have stated without proof, giving at best some informal arguments for believing them; rigorous proofs, which are elementary enough but moderately complicated, are, of course, available in the literature. Assuming these results for dimensions 3 and 4, the remainder of our discussion is rigorous and complete.

2. The standard regular solids

The tetrahedron, cube, and octahedron have analogs in all dimensions $n \geqslant 2$, and, as we have noted, these are the only regular n-solids, in E^n, for $n \geqslant 5$. We now describe these solids.

The n-simplex Δ_n. Our prototype, the tetrahedron Δ_3, has four vertices, with the distances between distinct pairs of vertices all the same; it is bounded by its four faces, each an equilateral triangular region (or Δ_2 in our present terminology) whose vertices are three of the four vertices of the tetrahedron. This description generalizes to give an inductive definition of Δ_n, assuming that Δ_{n-1} has been defined. Let $P_1,\ldots,P_{n+1}$ be points in E^n such that all the distances $d(P_i,P_j)$ for $i \neq j$ are equal. Then it is easy to see that each set of n of the P_i lies in an (n-1)-dimensional subspace, and hence can be taken as the set of vertices of a Δ_{n-1}. We take Δ_n to be the region in E^n bounded by these n+1 Δ_{n-1}'s.

[If one wants to be more concrete, one can choose the P_i to be the n+1 points $(0,\ldots,0,1,0,\ldots,0)$ in E^{n+1}, which clearly lie in the hyperplane with equation $x_1 + \ldots + x_{n+1} = 1$; this hyperplane can then be taken as E^n. One can define Δ_n as the convex closure of the P_i, that is, as the set of all points of the hyperplane such that all the coordinates x_i are nonnegative.]

That Δ_n is a regular solid follows from the fact that its automorphism group is given by the symmetric group of all permutations of the set of its n+1 vertices P_i.

Given an n-simplex Δ_n one can construct a dual solid Δ_n^*, and conclude that, since Δ_n is regular, Δ_n^* also is regular. Now Δ_n^* is defined to be the convex closure of its set of vertices, which are the centers of the faces of Δ_n. It is clear that Δ_n^* is simply another, smaller, n-complex, inscribed in Δ_n. Thus we obtain no new type of regular solid: the type Δ_n is selfdual.

The n-cube $\square_n$. Our prototype, the 3-cube $\square_3$, can be taken with the eight vertices $(\pm 1,\pm 1,\pm 1)$ in E^3. Each of the six planes $x_i = \pm 1$, for $i = 1, 2, 3$, contains four of these vertices, the vertices of a square (or $\square_2$). Evidently $\square_3$ is bounded by these six squares. For our inductive definition of $\square_n$, we have now 2^n vertices $(\pm 1,\ldots,\pm 1)$ in E^n, and each of the 2n planes $x_i = \pm 1$ contains 2^{n-1} of these vertices, which are the vertices of a $\square_{n-1}$. Now $\square_n$ is the region in E^n bounded by these 2n $\square_{n-1}$'s. [Alternatively, $\square_n$ is the set of all points $(x_1,\ldots,x_n)$ such that all $|x_i| \leqslant 1$.]

The dual $\square_n^*$ of the n-cube. This is harder to visualize, but easy enough to describe. We may take $\square_n^*$ with vertices the 2n centers of the 2n faces of $\square_n$. Let V be any vertex of $\square_n$, for simplicity take $V = (1,1,\ldots,1)$; the faces of $\square_n$ containing V are contained in the n planes $x_i = +1$, and their centers are the n points $(0,\dots,0,1,0,\dots,0)$. These n centers are among the vertices of $\Box_n^*$ and are the n vertices of a Δ_{n-1}. Now $\Box_n^*$ is bounded by 2^n such (n-1)-dimensional faces Δ_{n-1}, corresponding to the 2^n vertices of $\Box_n$. Except for $n = 2$, where $2^n = 2n$, $\Box_n^*$ is a regular solid of type different from $\Box_n$ and Δ_n.

Note. We do not really claim to visualize these solids, nor do we expect the reader to do so. It suffices to visualize the easy cases for $n = 2$ and $n = 3$ and to convince oneself that arguments carried out in these cases do not in fact depend on the value of $n \geqslant 2$.

Note. We have spoken in terms of regular solids, but we will be equally concerned with tessellations of higher dimensional spheres. If Π is a regular solid in E^n, and we choose its center at the origin 0, then all its vertices will be equidistant from 0, say at distance 1, and hence will lie on the (n-1)-sphere S^{n-1} with equation $x_1^2 + \dots + x_n^2 = 1$. If F is a face of Π. then projection from 0 maps F onto an (n-1)-dimensional piece F' of S^{n-1}. It is clear that since Π is a regular solid, the set of all these F' will form a regular tessellation T of S^{n-1}. Conversely, if a regular tessellation T of S^{n-1} is given, its vertices will be the vertices of a regular solid Π in E^n.

It results that the concept of regular solid and of regular tessellation of a sphere are essentially the same, and we will not always distinguish them carefully. However, it is important to note the shift in dimension: an n-dimensional regular solid in E^n corresponds to a regular tessellation of the (n-1)-dimensional sphere S^{n-1}. Indeed, it is this shift in dimension that makes possible the study of regular tessellations by induction on the dimension.

The standard tessellation K^n of E^n, $n \geqslant 1$, by n-cubes $\Box_n$, is that with vertices all points with integer coordinates.

3. Regular tessellations: examples

We have already considered regular tessellations of the sphere S^2 and of the Euclidean plane E^2. Here we required that the tiles be congruent regular (spherical or planar) p-gons for some $p \geq 3$, and that the same number $q \geq 3$ meet at each vertex. We now require a definition of a regular tessellation of the n-sphere S^n or of n-space E^n for $n > 2$. Before doing this, however, we want to broaden our definition of a tessellation by dropping all metric conditions, in accordance with our general goal of avoiding metric considerations as far as possible.

We will try to clarify this idea with three simple examples. The first is the tessellation T of the sphere S^2 of type (5,3), that is, the dodecahedron. To prove the existence of the metrically regular dodecahedron required an argument by continuity, or, alternatively, some trigonometry. But it is easy to prove the existence of a combinatorially regular dodecahedron, where it is required only that each region share a side with each of exactly 5 others and that there be exactly 3 regions at each vertex (end of a side). If the figure, as shown, is thought of as drawn on the sphere, then, starting with the central region, the construction of the successive rings as shown is forced at each step, and, with the figure as shown, the next step also is forced: to complete the tessellation with one more face, filling out the remainder of the sphere.

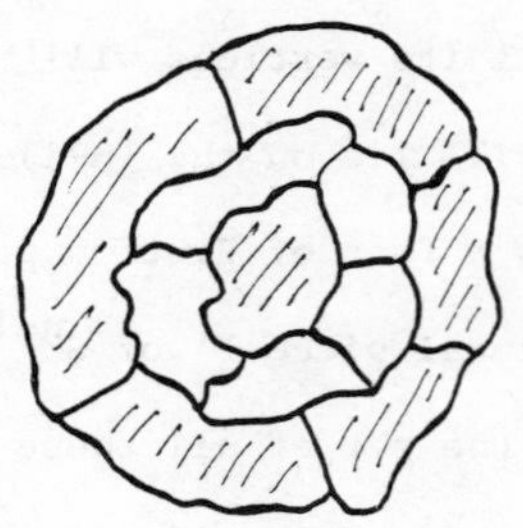

Our second example is the tessellation T of type (3,7) of the plane E^2. We require only that each region share a side with each of three others and that seven sides meet at each vertex. The beginning of the construction of such a tessellation T is shown in the figure, and we believe it is clear that the figure can be continued indefinitely and (provided the radii of the concentric circles are chosen to increase fast enough) will fill out the plane E^2 with a tessellation T of type (3,7). Now it is obvious that there is no tessellation T of type (3,7) of the plane in which the faces are congruent equilateral triangles in the Euclidean metric, although one is possible with faces that are congruent equilateral triangles in the hyperbolic metric. Our combinatorial approach does not distinguish between these two different metrics on the same topological space E^2.

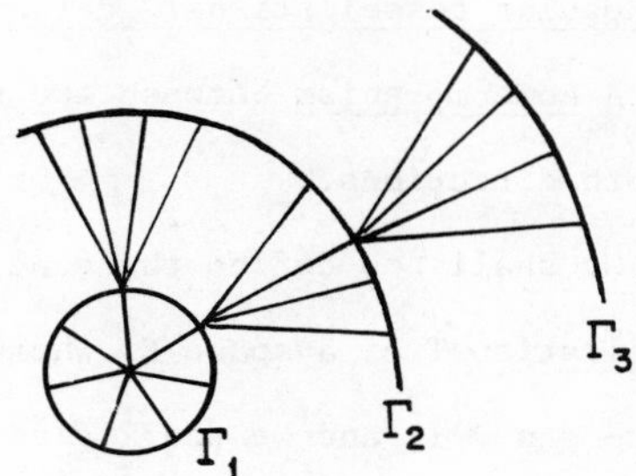

The third example is the tessellation of type (3,6) of E^2, which could be constructed as in the preceding example (if one didn't know better). But here the tessellation can be chosen regular in the Euclidean metric but not in the hyperbolic metric, the reverse of the case for (3,7).

Note. One may object that the combinatorial approach does in fact distinguish between the Euclidean and the hyperbolic metric, according as $1/p + 1/q$ is equal to or less than $1/2$, and presumably similar criteria exist for tessellations in higher dimensions. But our point is that we do not want to enter into the calculations necessary to establish such criteria.

4. Regular tessellations: definitions

A homeomorphism between two spaces is a bijection that is continuous in both directions.

We shall now define the concept of a combinatorially regular tessellation T of a space U, where U is homeomorphic to either S^n or E^n for some $n \geqslant 1$, and we shall also define the type of such a tessellation. For brevity we shall ordinarily omit the word 'combinatorially'.

Our definition will proceed by induction and must therefore begin with the case $n = 1$. If U^1 is homeomorphic to S^1, then U^1 is a simple closed curve, and a regular tessellation T of U^1 is a division of U^1 into p arcs by p vertices, for some $p \geqslant 3$; the type of T is p. If U^1 is homeomorphic to E^1, that is, a doubly infinite simple curve parametrized by E^1, a regular tessellation T of U^1 is a division of U into a countably infinite number of closed arcs by a countably infinite number of vertices; all such T have the same type K_1.

For our induction, let T be a tessellation of U^n, $n \geqslant 2$, that is, a division of U^n into nonoverlapping n-cells Π, each homeomorphic to the n-ball $B^n = \{(x_1,\ldots,x_n) : x_1^2 + \ldots + x_n^2 \leqslant 1\}$. Now $\partial\Pi \simeq S^{n-1}$ ($\partial\Pi$ is homeomorphic to S^{n-1}) and we require that there is a regular tessellation of $\partial\Pi$ with cells of the form $\Pi \cap \Pi'$ where Π' is another n-cell of the tessellation T. We require moreover that all these regular tessellations of the $\partial\Pi$ be of the same type X, which we term the face type of T.

Next, let V be any vertex of T, that is, a point that is the intersection of some set of cells Π; we assume that V is contained in only a finite number of cells Π, and we let W be the union of these cells. We assume that $W \simeq B^n$, whence $\partial W \simeq S^{n-1}$. Now ∂W is the nonoverlapping union of its intersections with the Π containing V, and we assume that this defines a regular tessellation of ∂W. We require moreover that all these regular tessellations of the ∂W, for all vertices V of T, are of the same type Y, which we term the vertex type of T. Finally, we define the type of T to be the ordered couple (X,Y).

Let us illustrate with the dodecahedral tessellation T of $U^2 = S^2$, described informally above. Each 2-cell Π is homeomorphic to B^2 (better known as the closed unit disc), whence $\partial\Pi \simeq S^1$, that is, $\partial\Pi$ is a simple closed curve. Now $\partial\Pi$ is the nonoverlapping union of 5 sides of the form $\Pi \cap \Pi'$, whence the face type is 5. Let V be any vertex of T. Then there are exactly 3 faces Π at V, and W is the union of these 3 faces. Thus $\partial W \simeq S^1$ is the nonoverlapping union of the three parts of ∂W contained in the three faces at V. We conclude that the vertex type of T is 3, and hence that T is of type (5,3).

As a second example, consider the tessellation T of E^3 into cubes by the planes $x_i = k$, for i = 1, 2, 3 ind $k \in \underline{Z}$. The face type is clearly (4,3), the type of the cube. The figure W at a vertex V is the union of the 8 cubes Π at V, but with the tessellation of ∂W one must be a little careful. It is that given by intersecting ∂W with each of the 8 cubes Π, thus a division of $\partial W \simeq S^2$ into 8 'triangular' regions, with 4 at each vertex. Thus the vertex type is that of the octahedran, (3,4). Since the face type is (4,3), we find that T is of type ((4,3),(3,4)).

We note that every tessellation T of U^n, homeomorphic to S^n or E^n, for $n \geqslant 2$, is of type (X,Y) where X and Y are types of regular n-solids, that is, of regular tessellations of S^{n-1}. We warn that, given an arbitrary pair of types X, Y of tessellations of S^{n-1}, there need not exist any tessellation of type (X,Y); a necessary and sufficient condition for this will be given presently.

A word now about metrically regular tessellations. A regular tessellation T of S^1 or E^1 is metrically regular if the arcs forming the tessellation are all of equal length. Inductively, using the same notation as above, a regular tessellation T is metrically regular if all the tessellations of the $\partial\Pi$ are metrically regular and congruent.

5. Existence

THEOREM. *Let* $X = (X_1, X_2)$ *and* $Y = (Y_1, Y_2)$ *be two types of regular tessellations of* S^{n-1}, $n \geq 3$. *Then there exist regular tessellations, of* S^n *or* E^n, *of type* (X,Y) *if and only if* $X_2 = Y_1$.

Example. We saw above that the tessellation of E^3 into cubes is of type ((4,3),(3,4)). In contrast, the theorem tells us that there is no regular tessellation of type ((4,3),(4,3)).

Proof. The proof of the necessity of the condition $X_2 = Y_1$ is essential to our goal of enumerating all regular tessellations, and will be given in full. The proof of the sufficiency of the condition comes essentially to constructing a regular tessellation of the given type (with $X_2 = Y_1$). Now we have already constructed all the regular tessellations in dimension n = 2, and the four standard types Δ_n, $\square_n$, $\square_n^*$, K_n in all dimensions, which, as it turns out, are all in dimensions $n \geq 5$. Thus there remains for consideration only dimension n = 3, where there are, in addition to the standard types, only one more selfdual type together with a dual pair (we shall construct these explicitly in Sections 9 and 10), and n = 4, where there is a dual pair of additional tessellations of E^4. Thus a sufficiency proof is in principle superfluous for our purposes, although perhaps still of methodological or philosophical interest. (It is hard to know how much importance to attach to a 'general theory' that in fact applies to only a finite number of objects.) For these reasons we shall say only a few informal words about sufficiency.

To prove necessity, assume then that T is a regular tessellation of type (X,Y) where $X = (X_1,X_2)$ and $Y = (Y_1,Y_2)$; we must show that $X_2 = Y_1$. The proof contains no new ideas, but is essentially a matter of tracing through the definitions.

Let V be a vertex of T and Π_0 an n-cell of T that contains V. As before, let $W \simeq B^n$ be the union of all n-cells Π of T that contain V, and let T_W be the tessellation of $\partial W \simeq S^{n-1}$ with (n-1)-cells all $F = \partial W \cap \Pi$ for n-cells Π of T that contain V. By definition, T_W is of type Y, and $F_0 = \partial W \cap \Pi_0$ is an (n-1)-solid of type Y_1, that is, $\partial F_0 \simeq S^{n-2}$ (as tessellated by T) is of type Y_1.

On the other hand, Π_0, as an n-cell of T, is an n-solid of type X, that is, $\partial\Pi_0 \simeq S^{n-1}$ is of type X. To calculate X_2, let $W_0 \simeq B^{n-1}$ be the union of all (n-1)-cells of $\partial\Pi_0$ that contain V, and let T_{W_0} be the tessellation of $\partial W_0 \simeq S^{n-2}$ by (n-2)-cells $\partial W_0 \cap P$ for all (n-1)-cells P of Π_0 that contain V. Then, by definition, T_{W_0} is of type X_2. Now $\partial\Pi_0 \simeq S^{n-1}$ is the union of two 'hemispherical' sets W_0, the union of all cells that contain V, and F_0, the union of the remaining cells. It follows that $\partial W_0 = \partial F_0$, whence $X_2 = Y_1$. This completes the proof of necessity.

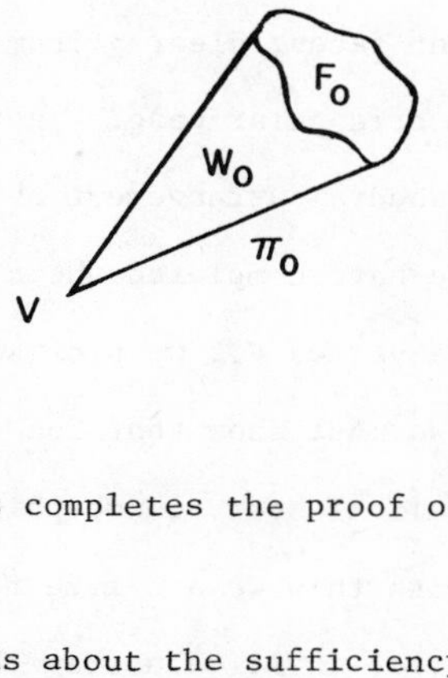

We will now say only a few informal words about the sufficiency of the condition $X_2 = Y_1$. Given types $X = (X_1,X_2)$ and $Y = (Y_1,Y_2)$ of n-solids, where $X_2 = Y_1$, we seek to construct a regular tessellation T of type (X,Y). Let us carry along as an example a transparent case where we already know the result: we take X = (4,3) and Y = (3,4); then T will turn out to be K_3, the regular tessellation of E^3 by cubes (or a homeomorphic replica of it).

We begin with a tessellation T_Y of S^{n-1} of type Y, here the projection of an octahedron on S^2. We divide B^n into 'sectors' $\hat{F}$ at its center 0, with 'bases' the faces F of T_Y. In our example, these sectors are pyramids at 0 with bases the triangular faces of the octahedron. We now 'hollow out' each of the $\hat{F}$, retaining only the 'lateral surface' $\tilde{F} = \partial\hat{F} - F$; these will serve as 'settings' into which we will insert corners of 'jewels' of shape X. The corner $\tilde{\Pi}$ of an n-cell Π of type X is defined to consist of all cells of $\partial\Pi$ at a fixed vertex V. Now the condition $X_2 = Y_1$ ensures that the $\tilde{F}$ and the $\tilde{\Pi}$ are isomorphic (but not necessarily metrically), so that we can identify each $\tilde{F}$ with the corner $\tilde{\Pi}$ of an n-cell Π (with the cell Π 'facing out'), to obtain a <u>cluster</u> C of n-cells Π at the common vertex 0, with vertex type Y (at 0). In the example, the corners consist of a vertex of a cube together with the three adjacent faces, clearly isomorphic to the lateral surface of the pyramid $\hat{F}$ with triangular base. In this example, we obtain a cluster of 8 cubes in octahedral arrangement about a common vertex 0.

We have completed the critical step. Now starting with a cluster T_1 about a vertex V_1, we pick some other vertex V_2 of T_1 (on the boundary of T_1). We must show that the set of cells Π <u>of T_1</u> at V_2 can be completed to a full cluster, isomorphic to C, by adding new cells to form a figure T_2. With this we are free to iterate the construction, obtaining $T_3, T_4, \ldots$. At each stage T_k will be homeomorphic to B^n, unless at some stage we encounter T_k without boundary, that is, $T_k \simeq S^n$, in which case we are done, with $T = T_k$. If this does not happen (and if we proceed judiciously) we have an infinite ascending chain of figures T_k, whose union is a regular tessellation T, of required type, of a space U homeomorphic to E^n.

If this is a little too vague, the reader is referred to the concrete examples in Sections 9 and 10.

6. Duality

The concept of duality is, like the preceding existence proof, not essential to our purpose, but is instructive and useful. The situation is similar in that we have already discussed duality in dimension $n = 2$ and also for the standard tessellations. Thus, to complete the discussion of duality for the regular tessellations of spheres, only three cases in dimension 3 remain. So, once again, we content ourselves with a sketch; more precisely, we give definitions and results for the case of a metrically regular tessellation, leaving it to the interested reader to adjust them to the case where metric regularity is not assumed.

Let T be a metrically regular tessellation of $U^n = S^n$ or E^n, $n \geqslant 2$. To define a dual tessellation T^* of the same space U^n, we begin by specifying its vertices as the centers Π^* of the n-cells Π of T. Next, T^* will have an n-cell V^* corresponding to each vertex V of T; the vertices of V^* are precisely the Π^* corresponding to n-cells Π of T containing V. This fully determines V^* (e.g., as the _convex closure_ of the set of its vertices, obtained by repeatedly adjoining all points lying on segments (of lines or great circles) between points already present). From this definition it is easy to see that $T^{**} = T$, hence the term 'dual' is fully justified.

There is a great deal more that could be said about duality, but we confine ourselves to one result that is relevant to our purposes.

THEOREM. _If_ T _is a regular tessellation of type_ (X,Y), _and_ T^* _is a regular tessellation dual to_ T, _then_ T^* _is of type_ (Y^*,X^*).

Proof. We shall use arguments that are clearly valid in the metrically regular case, and which remain valid under a proper extension of the concept of duality to the general case. Let T be a regular tessellation of U^n, $n \geqslant 2$, of type (X,Y). Let V be a vertex of T and W the union of all n-cells Π of T that contain V. Then, by definition, the tessellation T_Y of ∂W into faces $F = \partial W \cap \Pi$, for all Π in T that contain V, is of type Y.

Let V^* be the n-cell of T^* corresponding to the vertex V of T. Then V^* is a regular n-solid of some type Z, that is, T^* induces on $\partial V^* \simeq S^{n-1}$ a tessellation T_V^* of some type Z, and we must show that $Z = Y^*$.

Now V^* has vertices Π^*, one contained in each n-cell Π of T that contains V, and V^* is contained within W. We now proceed to 'project' ∂V^* outward onto ∂W. First, we project each vertex Π^* of V^* out onto a vertex F^* contained in $F = \partial W \cap \Pi$, and then we project each cell of ∂V^* with vertices $\Pi_1^*,\ldots,\Pi_k^*$ onto a corresponding cell on ∂W with vertices $F_1^*,\ldots,F_k^*$. Without entering into details we may say that the same conditions that make V^* a cell in the dual T^* of T serve to make the resulting projection of ∂V^* onto ∂W dual to T_Y. We conclude that ∂V^*, as tessellated by T^*, is of the type Y^* dual to Y, that is, $Z = Y^*$ as required.

We have shown that T^* is of type (Y^*,Q) for some Q. But, since $T^{**} = T$ is of type (X,Y), we conclude by the same argument that $X = Q^*$, whence $Q = X^*$. This completes the proof that T^* is of type (Y^*,X^*). □

7. Regular tessellations of dimensions 2 and 3

We are at last ready to begin our enumeration of the regular tessellations of S^n and E^n for $n \geqslant 2$. We begin by recapitulating the information we already have for the case n = 2. We present this information in a table, which we now explain. The entry in row p and column q is the regular tessellation of type (p,q). The entries Δ, $\Box$, $\Box^*$, and K represent $\partial\Delta_3$, $\partial\Box_3$, $\partial\Box_3^*$, and K_2, that is, the tetrahedral, cubical, octahedral tessellations of S^2, and the regular tessellation K_2 of E^2 by squares. (We shall henceforth systematically suppress the symbol ∂ and the subscripts indicating dimension, which are clear from the context.) The entry P denotes the regular dodecahedral tessellation of S^2, and P^* the dual icosahedral tessellation. T denotes the regular tessellation of E^2 by triangles, 6 at a vertex, and T^* the dual tessellation by hexagons. The remaining entries (some marked with .'s) represent types (p,q) of regular tessellations of E^2 that are not metrically realizable in the Euclidean metric, but only in the hyperbolic metric.

p \ q	3	4	5	6	7	...
3	Δ	$\Box^*$	P^*	T	.	
4	$\Box$	K	.	.	.	
5	P	.	.	.	.	
6	T^*	.	.	.	.	
7	.	.	.	.	.	
⋮						

We now turn to a similar tabulation of the possible types of regular tessellations of S^3 or E^3. Such a type must have the form (X,Y) where X and Y are types of regular tessellations of S^2 (or regular 3-solids); since there are only 5 such types, our table will be finite.

In this table the x's mark pairs (X,Y) which are not the type of any regular tessellation, since the condition $X_2 = Y_1$ fails. Of the remaining entries, Δ, $\Box$, and $\Box^*$ denote the standard tessellations of S^3, and K the standard tessellation of E^3. The entry A denotes a selfdual finite tessellation with 24 3-cells of S^3. Its existence as a (combinatorial) regular tessellation will be established in Section 9. It is well known, but will not be proved here, that A can in fact be realized as a metrically regular tessellation of S^3, thus giving rise to a 24-sided regular 4-solid in E^4. Similarly, the entry B denotes a finite tessellation with 600 3-cells of S^3; we establish its existence as a (combinatorial) regular tessellation in Section 10, but not the well known fact that it can be realized as a metrically regular tessellation of S^3. The entry B^* denotes the dual of B, a regular tessellation with 120 3-cells of S^3.

	△	□	□*	P	P*
△	△	x	□*	x	B
□	□	x	K	x	C
□*	x	A	x	x	x
P	B*	x	C*	x	D
P*	x	x	x	E	x

The remaining entries C, C^*, D, E represent CRT's of E^3, but it is known that they are not realizable as MRT's of E^3 in the Euclidean metric, but are metrically realizable in the hyperbolic metric. (For this latter point see Coxeter, Twelve Geometric Essays, Chapter 10.)

8. Regular tessellations in dimensions $n \geqslant 4$

We have established the existence of tessellations of S^3 of types △, □, □*, A, B, and B^*. To continue, we must assume what was not completely proved, that these are all. Using these six entries as labels for the rows and columns, we now make a table of the regular tessellations of dimension n = 4. In this table the x's again mark pairs that are not the type of any regular tessellation. The entries

	△	□	□*	A	B	B*
△	△	x	□*	x	F	x
□	□	x	K	x	G	x
□*	x	x	x	H	x	x
A	x	H*	x	x	x	x
B	x	x	x	x	x	x
B*	F*	x	G*	x	x	J

△, □, □*, K represent standard tessellations. The remaining entries represent regular tessellations of E^4. The types H, H^* are known to be realizable by metrically regular tessellations of E^4 in the Euclidean metric, although we shall not prove it, and the remaining types F, F^*, G, G^*, and J are not metrically realizable in the Euclidean metric but are in the hyperbolic metric.

Again, to continue we must assume that the three standard types Δ, $\square$, $\square^*$ are the only types of regular tessellation of S^4. Then the table for regular tessellations of dimension n = 5 is as shown, and we conclude directly that the three standard types are the only types of regular tessellation of S^n for all $n \geqslant 4$, and also that the standard tessellation K is the only type of regular tessellation of E^n for all $n \geqslant 5$.

	Δ	$\square$	$\square^*$
Δ	Δ	x	$\square^*$
$\square$	$\square$	x	K
$\square^*$	x	x	x

We summarize these results, not all of which have been proved, in a table listing the types of regular tessellation of spherical, Euclidean, and hyperbolic space for all dimensions $n \geqslant 2$.

	Spherical	Euclidean	Hyperbolic
n = 2	Σ, $\square$, $\square^*$, Π, Π^*	K, L, L^*	all (p,q), $\frac{1}{p}+\frac{1}{q}<\frac{1}{2}$
n = 3	Σ, $\square$, $\square^*$, A, B, B^*	K	C, C^*, D, E
n = 4	Σ, $\square$, $\square^*$	K, H, H^*	F, F^*, G, G^*, J
n = 5	Σ, $\square$, $\square^*$	K	none

(We note that Coxeter uses a broader definition of hyperbolic tessellation than we have, and consequently lists more hyperbolic types than appear in our table; see the second paragraph of his essay cited above.)

All the assertions contained in the table for n = 2 have been proved. The existence of metrically regular tessellations of types Σ, $\square$, $\square^*$, and K in all dimensions has been proved. The existence of combinatorially regular tessellations of S^3 of types A, B, B^* will be proved in

Sections 9 and 10, but we do not prove the existence of metrically regular tessellations of these types. The existence of metrically regular Euclidean tessellations of E^4 of types H, H^* is not proved; we do not even prove that this tessellation is infinite. We have not proved that the types C, C^*, D, E in dimension n = 3 and the types F, F^*, G, G^*, J in dimension n = 4 are infinite, that they are not realizable as regular Euclidean tessellations of E^n, or that they are in fact realizable as regular tessellations of hyperbolic space.

Some informal suggestions for proving the unproved assertions above will be given in Section 11 and in the Problems.

9. The regular 4-dimensional solid with 24 cells

We shall now show that the regular tessellation T of type $A = (\square_3^*, \square_3)$ is finite, hence a regular tessellation of S^3, giving rise to a regular solid in E^4. We shall not show that A can be realized by a metrically regular tessellation. We note that A is selfdual, $A^* = A$.

The cells of T are to be octahedral solids Π with 6 at each vertex, meeting in the figure of a cube. We follow the general procedure outlined earlier, starting with a cluster T_1 of 6 cells at a single vertex, and extending T_1 successively to larger figures T_k, all homeomorphic to B^3 until the last stage, where the construction 'closes up' to yield a T_k homeomorphic to S^3. Although this is a 3-dimensional construction, we can reduce it to 2-dimensional considerations by the observation that, to know how new cells can be added, it suffices to know ∂T_k together with the number of cells of T_k incident with each vertex and edge of ∂T_k. (Of course, a (triangular) face of ∂T_k will be incident with exactly one cell of T_k.)

Instead of adding one cell at a time, or of completing a single cluster, we exploit the natural symmetry of T_1 by adding new cells symmetrically at each stage in such a way that T_k retains this symmetry. We do not have to continue this process very long: it turns out that we are already done with $T_3 \simeq S^3$. One might describe $T = T_3$ as follows: T_1 is a polar cap, say at the north pole; we form T_2 by surrounding T_1 with an equatorial belt, and then find that a second polar cap, isomorphic to T_1, at the south pole suffices to complete the tessellation $T = T_3$ of S_3.

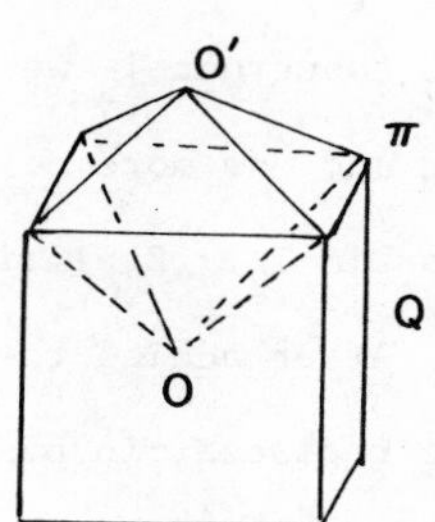

To construct T_1 we start with a cube Q with center O. If F is a face of Q, let O' be the image of O reflected in F. Then O, O' and the 4 vertices of the face F are the six vertices of a (combinatorial) octahedron Π. The six such octahedra Π obtained thus from the 6 faces F of Q evidently provide our initial cluster T_1.

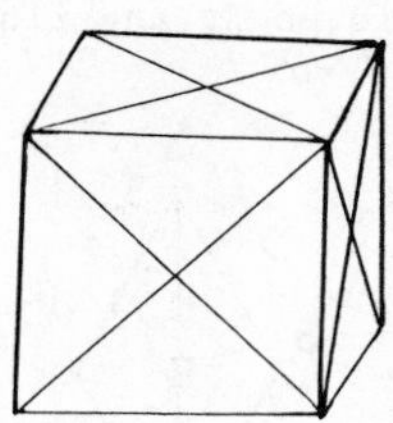

Evidently ∂T_1 is isomorphic to the tessellation obtained from ∂Q by cutting each face F of Q into 4 parts by its two diagonals. Clearly T_1 retains the symmetry of Q, whence, to describe it, it more than suffices to show the part corresponding to a single face of Q together with adjoining cells. In our (partial) sketch of ∂T_1, we have marked certain

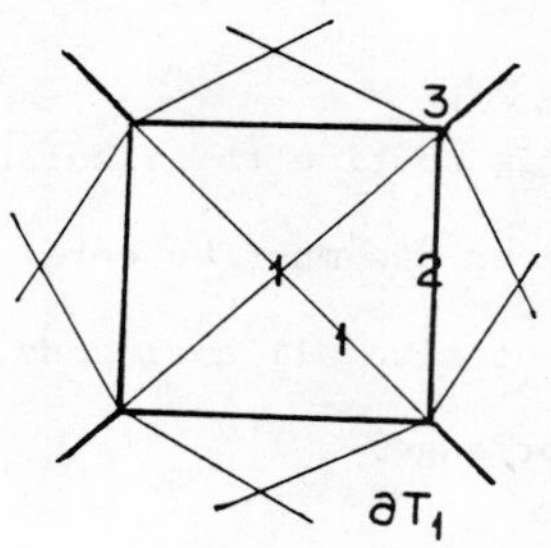

vertices and edges with marks indicating the number of cells Π of T_1 with which they are incident. We have exploited the obvious symmetries in not cluttering the figure with redundant markings.

Before we can continue, we must know the expected number of cells Π at a vertex or edge in the completed tessellation T; at any stage, a vertex or edge will become interior when it has acquired this full quota. This is easily settled, since the interior vertex 0 of T_1 is on exactly 6 cells, while an interior edge, from 0 to a vertex of Q, lies on exactly 3 cells.

To construct T_2 we note that ∂T_1 contains edges with mark 2; we propose to add one more cell at each such edge, so that the edge becomes interior in T_2. Explicitly, we isolate a part of the sort P pictured below. After adding the new cell Π (and analogous new cells) the part P will be replaced, in passing to ∂T_2, by a new part P'. In the figure we show first P, then a schematic sketch of the new cell Π indicating how P and P' lie on Π, and finally the resulting figure P'.

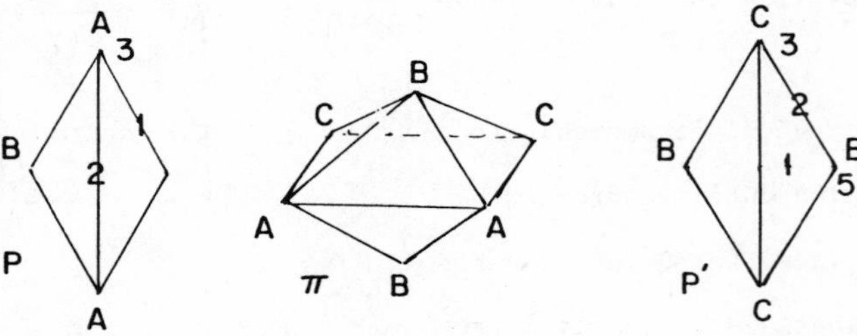

It is routine to calculate the marks on the vertices and edges of P'; however one must be careful to take into account not only the new cell Π but also all analogous new cells that may be incident with the vertex or edge.

We note that ∂T_2 has the same form as ∂T_1, except that in passing from ∂T_1 to ∂T_2 the marks on corresponding vertices and edges have been changed. We note further that the sum of the marks on corresponding vertices of ∂T_1 and ∂T_2 is always 6, a full quota, and that the sum of the marks on corresponding edges is always 3, again a full quota. From this it follows, as indicated earlier, that if we unite T_2 and a replica T_1' of T_1 along their boundaries according to this correspondence, all vertices and edges (and therefore also all faces) will have become interior, to give us $T = T_3 = T_2 \cup T_1'$ homeomorphic to S^3.

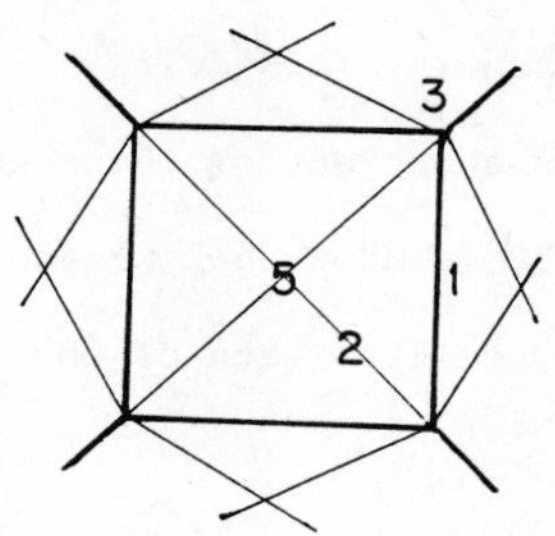

To count the number of cells Π in T we need only note that T_1 had 6 cells and that 12 more were added in passing to T_2, whence T has in all 6 + 12 + 6 = 24 cells. Since A is selfdual, we conclude directly that T has also 24 vertices, and the same number of edges as faces. In fact, since there are 2 vertices on each edge and 8 edges at each vertex, counting in two ways the number of pairs consisting of an incident vertex and edge we conclude that there are 4 times as many edges as vertices, hence 96 edges and also 96 faces.

10. The regular 4-dimensional solid with 600 cells

We now construct a finite tessellation T of type $B = (\Delta_3, P^*)$, with 600 tetrahedral cells and icosahedral vertex type. Since the procedure here is entirely parallel to that given for type A above, we exhibit the various stages of the construction with a minimum of explanation.

The initial cluster T_1 is obtained by dividing a regular icosahedral solid into 20 tetrahedral cells Π, with the 20 faces of the icosahedron as bases and a common vertex at the center 0. We observe that there is a full quota of 20 cells at the central vertex 0, and a full quota of 5 cells about any of the edges at 0, interior to T_1. A representative part of ∂T_1 is shown below, and also of ∂T_2, where T_2 is obtained by attaching a new cell to each of the triangular faces of ∂T_1.

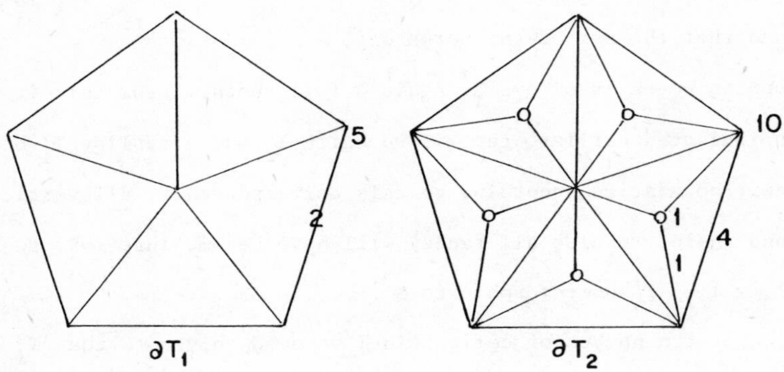

To pass from T_2 to T_3 we attach one new cell at each edge of ∂T_2 with mark 4, whereupon this edge becomes interior to T_3. Thus, in passing from ∂T_2 to ∂T_3, each part of the form P_2 is replaced by one of the form P_2', where P_2 and P_2' are as shown.

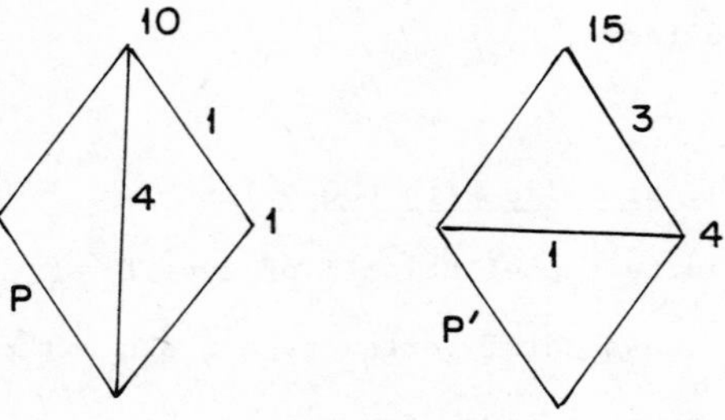

The figures for ∂T_3 and ∂T_4 are as shown below.

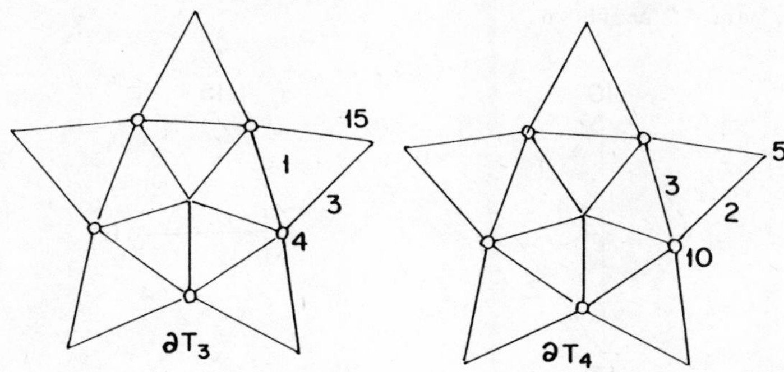

The passage from T_3 to T_4 is accomplished by attaching 5 new cells to the 5 faces of each part of ∂T_3 of the form P_3, which is then replaced in T_4 by a part of the form P_3', where P_3 and P_3' are as shown below.

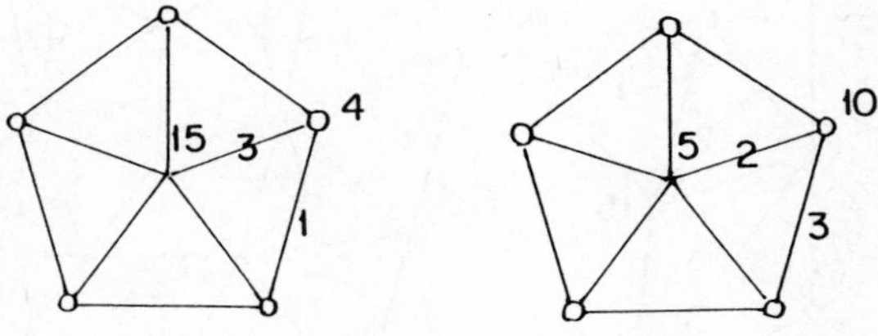

To pass from T_4 to T_5, we add 2 new cells to each part P_4 to obtain a part P_4' as shown.

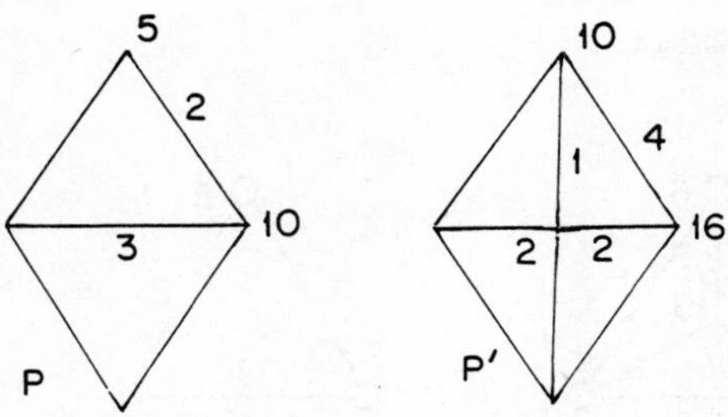

To pass from T_5 to T_6, we add a single cell to each part P_5 to obtain a part P_5' as shown.

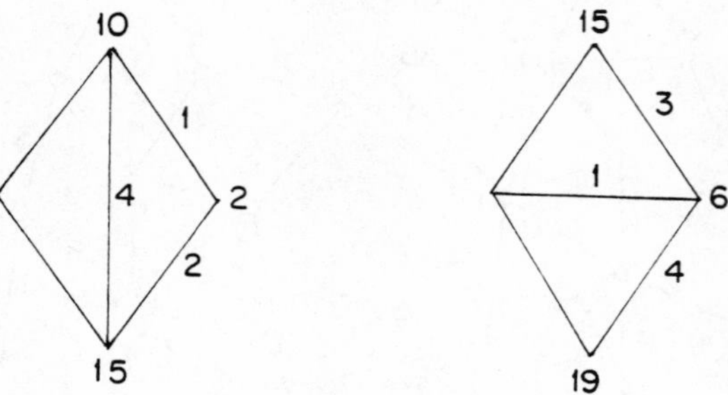

The resulting figures ∂T_5 and ∂T_6 are shown below.

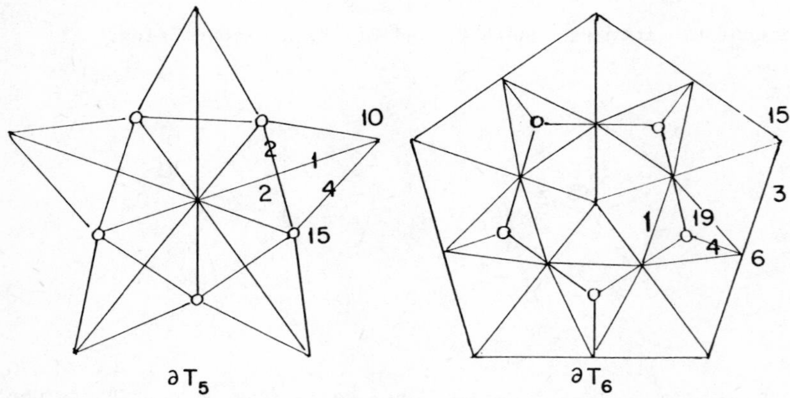

∂T_5 ∂T_6

Next, we pass from T_6 to T_7 by attaching one new cell to each part P_6 to obtain P_6', as shown.

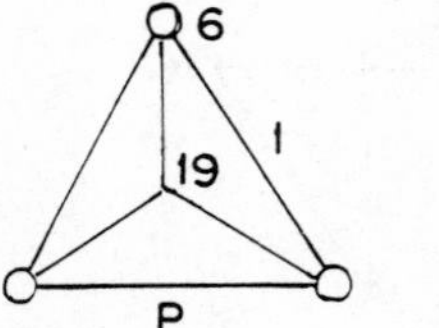

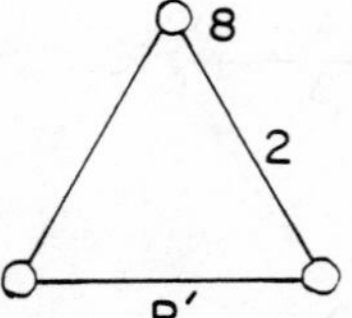

This yields ∂T_7 as shown below.

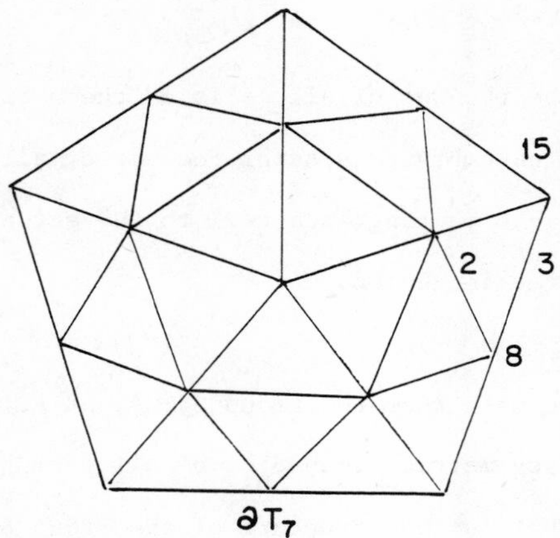

Finally we pass from T_7 to T_8 by attaching 5 new cells to each part P_7, to obtain a part P'_7, as shown.

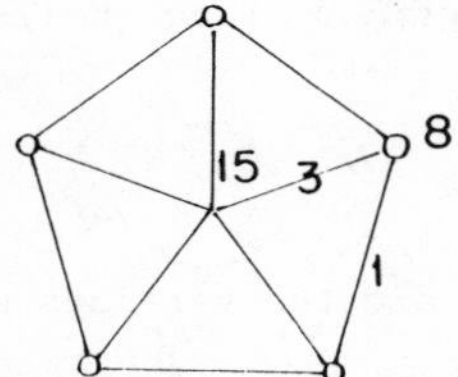

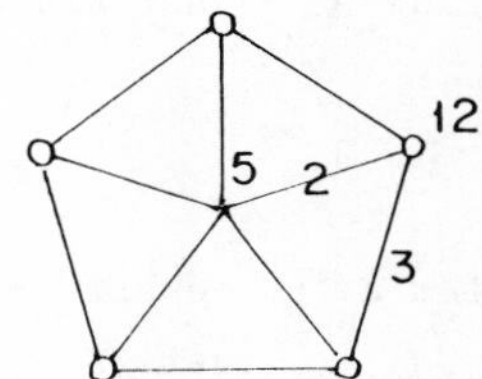

At last we observe that resulting ∂T_8 has the same form as ∂T_7 except for the markings, and that, under the obvious correspondence between ∂T_7 and ∂T_8, the sum of the marks on corresponding vertices is 20 and the sum of the marks on corresponding edges is 5. As in the case of A, treated in the last section, we can now unite T_8 with a replica T'_7 of T_7 by identifying ∂T_8 and $\partial T'_7$ according to this correspondence to obtain the desired tessellation T of S^3 of type B.

A count shows that T has $a_3 = 600$ tetrahedral cells, $a_2 = 1200$ triangular faces, $a_1 = 720$ edges, and $a_0 = 120$ vertices. In particular, we conclude that the dual type B^* is that of a combinatorially regular tessellation of S^3 by 120 dodecahedral solids, 4 at each vertex.

Problems

Problem 1. Let C be the set of all cells of the n-simplex Δ_n, and let D be the set of all nonempty subsets of the set of all vertices of Δ_n. Then the map $f : C \to D$ mapping each cell to the set of its vertices is bijective and preserves inclusion.

Problem 2. We have seen that the group Sym Δ_n of all symmetries of Δ_n is isomorphic to the symmetric group S_{n+1} of all permutations of the set of its vertices. What is the structure of the group Sym $\square_n$?

Problem 3. For $\square_n$ with vertices $V = (\pm 1,\ldots,\pm 1)$, what are the vertices of the inscribed $\square_n^*$? What are the vertices of an (n-1)-dimensional face of $\square_n^*$?

Problem 4. For K_n as described, what are the vertices of K_n^* and what are its n-cells? If S^{n-1} is the sphere at the origin with radius $r < 1$, what is the tessellation T_0 of S^{n-1} defined by intersecting S^{n-1} with the tessellation K_n?

Problem 5. A type of regular 3-tessellation is of the form $C = ((p,q),(q,r))$, which we abbreviate to $C = (p,q,r)$. A type of regular 4-tessellation is of the form $C = ((p,q,r),(q,r,s))$ which we abbreviate to $C = (p,q,r,s)$. Extending this notation in the obvious way, for what integers $p_1,\ldots,p_n$ does there exist a CRT of type $C = (p_1,\ldots,p_n)$? What is the dual type?

Problem 6. A nontrivial rotation σ in E^3 must fix all points of a line ℓ and leave invariant every plane P orthogonal to ℓ. If a translation τ is not parallel to ℓ, then it does not leave ℓ invariant whence $\sigma^\tau \neq \sigma$. Since the axis $\ell\tau$ of σ^τ is parallel to ℓ, both σ and σ^τ, and so also the translation $\tau_1 = \sigma^{-1}\sigma^\tau$ leave P invariant.

If σ and τ are contained in a discontinuous group G of isometries of E^3, then the subgroup generated by σ and τ_1 acts as a discontinuous group of isometries of the plane P. Use this argument to conclude that G cannot contain rotations of orders other than 1, 2, 3, 4, and 6, and that there can be no MRT of E^3 with dodecahedral or icosahedral tiles.

Problem 7. In a certain finite society, each boy has met exactly p girls and each girl has met exactly $q \geqslant 1$ boys. Show that there are p/q times as many boys as girls.

Problem 8. A graph Γ in the plane is a set of simple arcs, or edges, such that two different edges meet at most in 1 or 2 common endpoints. Suppose that Γ is nonempty, finite and connected. Let a_0 be the number of endpoints (vertices), a_1 the number of edges, and a_2 the number of bounded components of the complement of Γ in the plane (faces).
Probe that $\chi(\Gamma) = a_0 - a_1 + a_2 = 1$.
(Hint: For $a_2 = 0$, use induction on a_1, and then use induction on a_2.)

Problem 9. This is a rather extensive problem: For a fuller discussion see Coxeter, Regular Polytopes, Chapter 9.

If T is any finite 'complex' consisting of k-cells of dimensions $k = 0, 1, \ldots, n$ in Euclidean space, then the Euler-Poincaré characteristic $\chi(T)$ of T is defined to be

$$\chi(T) = a_0 - a_1 + a_1 - \dots + (-1)^n a_n ,$$

where, for each k, a_k is the number of k-cells. It is known that if T is a (finite) regular tessellation of S^n, $n \geq 1$, and indeed under considerably more general hypotheses, that $\chi(T) = 1 + (-1)^n$, that is, $\chi(T) = 2$ if n is even and $\chi(T) = 0$ if n is odd. The case n = 2, where T is a tessellation of the 2-sphere S^2 (not required to be regular) is the celebrated theorem of Descartes and Euler.

If we define $c_k = \frac{a_k}{a_0}$ for $k = 0, 1, \dots, n$, then $\chi(T) = \theta(T) \cdot a_0$ where

$$\theta(T) = c_0 - c_1 + c_2 = \dots + (-1)^n c_n .$$

Thus, if n is even $\theta(T) = \frac{2}{a_0}$ while if n is odd $\theta(T) = 0$. Now, counting pairs consisting of a k-cell and incident vertex in two ways, one sees that $c_k = \frac{q_k}{p_k}$ where p_k is the number of vertices on a k-cell and q_k is the number of k-cells at a vertex. If the type (X,Y) of a regular tessellation T of S^n is given, it is routine if tedious to calculate the p_k and q_k, and from them the c_k and so $\theta(T)$.

The first, elementary, part of our exercise is to carry out these calculations for all regular tessellations of S^n, $n \geq 1$, and thus verify the Euler-Poincaré formula.

We now observe that given any type (X,Y) of a regular tessellation T, without knowing whether T is finite or not, we can carry out the same calculations to determine $\theta(T)$. If n is odd and $\theta(T) \neq 0$, we conclude that T is not a tessellation of S^n. If n is even and $\theta(t)$ is not of the form $\frac{2}{v}$ for v a reasonably large positive integer, we conclude again that T is not a tessellation of S^n.

The second part of our exercise is to apply this test to all the regular tessellations in dimensions n = 2 and n = 3. If our calculations are correct, this test yields nothing for $n = 3$, where $\theta(T) = 0$ in all cases, but, for $n = 4$, it excludes all but the known regular tessellations of S^4.

The third part of our exercise is vaguer, to rectify, or refute the following rough argument. Suppose that T is a metrically regular tessellation of E^n (we have in mind here only $n = 4$) and T_r, for $r > 0$, is the part of T consisting of all cells contained within a ball B_r of radius r at the origin. Since T_k together with one complementary n-cell constitutes an irregular tessellation of S^n, we are inclined to believe that $\chi(T_r) = [1 + (-1)^n] - (-1)^n = 1$. Let $\theta(T)$ be calculated as before, and let $a_0(r)$ be the number of vertices of T_r. Then $\theta(T)a_0(r)$ is only an approximation to $\chi(T_r)$, because of irregularities at vertices on the boundary ∂T_r. If $b(r)$ is the number of vertices on ∂T_r, then $\lim \frac{b(r)}{a_0(r)} = 0$ as r increases indefinitely, while the error is of the form

$$|\chi(T_r) - \theta(T)\cdot a_0(r)| \leqslant N\cdot b(r)$$

for some constant $N > 0$. Dividing by $a_0(r)$ and passing to the limit we conclude that $\theta(T) = 0$. Granted this rather tenuous argument we conclude that T can be a metrically regular tessellation of E^n only if $\theta(T) = 0$. If our rather precarious calculations are not at fault, this shows that K_4 and H, H^* are indeed the only metrically regular tessellation of the Euclidean space E^4.

References.

A very careful discussion of the geometry of four dimensions, and its history up to the beginning of this century, is given by Manning. A very broad discussion, with an abundance of historical remarks and of further references, of the subject of this chapter (among other things), is contained in Coxeter's Regular Polytopes. Both books repay even a cursory examination.

CHAPTER SIX: INCIDENCE GEOMETRY OF THE AFFINE PLANE

1. Combinatorial description of the affine group

The affine group A was defined with reference to the metric in the Euclidean plane E, and was described in terms of real matrices. However, the affine group does not preserve the Euclidean metric, and it is possible to give a characterization of A that makes no reference to this metric. We introduce the temporary notation L for the group of all bijections from E to E that map lines to lines. We shall show that, in fact, $A = L$.

We outline the proof before turning to details. It is clear that $A \subseteq L$, whence it remains only to prove that $L \subseteq A$. Now this inclusion reduces easily to the assertion that if α in L fixes two distinct points of a line ℓ, then α fixes all points of the line. Introducing coordinates, we may suppose that the line ℓ is the x-axis and that α fixes (0,0) and (1,0). It is clear that α maps ℓ bijectively to itself, whence the equation $(x,0)\alpha = (x\gamma,0)$ defines a bijection γ of $\underline{R}$. The next step is to show that γ is an automorphism of $\underline{R}$, $\gamma \in \text{Aut } \underline{R}$, in the sense that, for all $x, y \in \underline{R}$, $(x + y)\gamma = x\gamma + y\gamma$ and $(xy)\gamma = (x\gamma)(y\gamma)$. The final step is to show that the identity is the only automorphism of $\underline{R}$, that is, if $\gamma \in \text{Aut } \underline{R}$, then $x\gamma = x$ for all $x \in \underline{R}$.

As is not uncommon, the formal proof runs in the opposite order to the informal outline. We begin with the following.

THEOREM. Aut $\underline{R} = 1$.

Proof. We assume that γ is a bijection from $\underline{R}$ to $\underline{R}$ such that $(x + y)\gamma = x\gamma + y\gamma$ and $(xy)\gamma = (x\gamma)(y\gamma)$ for all $x, y \in \underline{R}$. We must show that $x\gamma = x$ for all $x \in \underline{R}$.

Since $0 + 0 = 0$, we must have $0\gamma + 0\gamma = 0\gamma$, which implies that $0\gamma = 0$. Since $1 \cdot 1 = 1$, we must have $(1\gamma)(1\gamma) = 1\gamma$ moreover, $1 \neq 0$ implies that $1\gamma \neq 0\gamma = 0$, whence we may conclude that $1\gamma = 1$. Since $(1 + 1 + \ldots + 1)\gamma = 1\gamma + 1\gamma + \ldots + 1\gamma = 1 + 1 + \ldots + 1$, for a sum of n terms 1, it follows that $n\gamma = n$ for all positive integers n. From $n + (-n) = 0$ it follows that $n\gamma + (-n)\gamma = 0\gamma = 0$, whence $(-n)\gamma = -(n\gamma)$, and we conclude that $n\gamma = n$ for all $n \in \underline{Z}$. Similarly, if $x = \frac{m}{n} \in \underline{Q}$, with $m, n \in \underline{Z}$ and $n \neq 0$, then $nx = m$, $(n\gamma)(x\gamma) = (m\gamma)$, that is, $n(x\gamma) = m$, and we conclude that $x\gamma = x$. Thus $x\gamma = x$ for all $x \in \underline{Q}$.

Next we observe that $x \geqslant 0$ if and only if $x = y^2$ for some $y \in \underline{R}$, hence if and only if $x\gamma = (y\gamma)^2$ for some $y\gamma$, that is, if and only if $x\gamma \geqslant 0$. Now $x \leqslant y$ if and only if $y - x \geqslant 0$, equivalent to $y\gamma - x\gamma \geqslant 0$, and to $x\gamma \leqslant y\gamma$. Thus γ preserves the order of $\underline{R}$.

We now use the fact that $\underline{Q}$ is dense in $\underline{R}$, in the sense that if $x, y \in \underline{R}$ and $x < y$, then there is some $z \in \underline{Q}$ between them, with $x < z < y$. Finally, for all $x \in \underline{R}$ we define $L_x = \{y \in \underline{Q} : y \leqslant x\}$; it follows from the density of $\underline{Q}$ in $\underline{R}$ that $L_x = L_y$ if and only if $x = y$. Now, L since $y\gamma \leqslant x\gamma$ if and only if $y \leqslant x$, $L_{x\gamma} = \{y \in \underline{Q} : y\gamma \leqslant x\gamma\} = \{y \in \underline{Q} : y \leqslant x\} = L_x$, whence $x\gamma = x$. We have shown that $x\gamma = x$ for all $x \in \underline{R}$, that is, $\gamma = 1$, the identity automorphism of $\underline{R}$. □

We next describe simple geometric constructions for addition and multiplication, that is, given points $(x,0)$ and $(y,0)$ in a certain coordinate system, constructions for the points $(x + y,0)$ and $(xy,0)$.

We begin with addition. Let ℓ be a line and 0 a point of ℓ. We give a construction which, given points A and B of ℓ, yields a point C of ℓ such that $\overrightarrow{OA} + \overrightarrow{OB} = \overrightarrow{OC}$. The construction is shown in the figure. It begins with the choice of an arbitrary point P not on ℓ, through which are drawn the lines OP, PB, and ℓ', the parallel to ℓ through P. Next the parallel m to OP through A is drawn. Since P is not on ℓ, OP is not parallel to ℓ, whence m is not parallel to ℓ', and m and ℓ' meet in a point Q. Next the parallel n to PB through Q is drawn. We confirm that n is not parallel to ℓ, whence n and ℓ meet in a point C. Now, using the fact that opposite sides of parallelograms are equal, we find that $\overrightarrow{OC} = \overrightarrow{OB} + \overrightarrow{BC} = \overrightarrow{OB} + \overrightarrow{PQ} = \overrightarrow{OB} + \overrightarrow{OA}$. Note that although the construction requires a point P not on ℓ, the result does not depend on the particular point P chosen. We introduce the notation $C = A \oplus B$, where it should be noted that the operation $\oplus$, defined on ℓ, depends on the choice of the point 0.

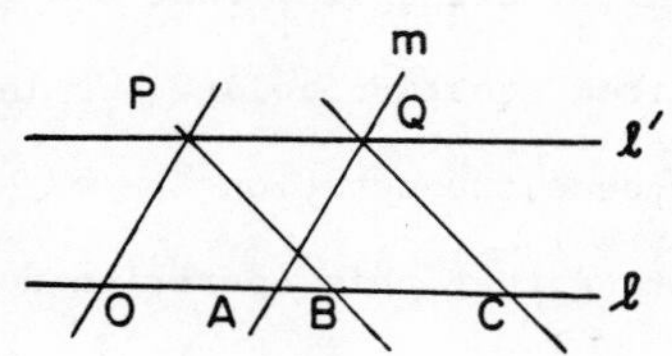

We record our first modest result.

LEMMA. <u>Let</u> $\alpha \in \ell$ <u>leave the line</u> ℓ <u>invariant and fix the point</u> 0 <u>of</u> ℓ. <u>Then, for all</u> A, B <u>on</u> ℓ, $(A \oplus B)\alpha = A\alpha \oplus B\alpha$.

<u>Proof</u>. The transformation α carries the figure used to construct $C = A \oplus B$ into a figure that constructs $C\alpha = A\alpha \oplus B\alpha$. □

We now turn to multiplication. We now suppose that a line ℓ and two distinct points 0 and I of ℓ are given. Our goal is to give a construction which, given two points A and B on ℓ, yields a third point C such that $\overrightarrow{OC} = \dfrac{\overrightarrow{OA} \cdot \overrightarrow{OB}}{\overrightarrow{OI}}$. We again choose an arbitrary point P not on ℓ, draw the lines OP, IP, AP, then BQ parallel to IP through B and meeting OP in Q, and finally QC parallel to AP through Q, meeting ℓ in C. We leave it to the reader to verify, using similar triangles, that the required equation holds. We introduce the notation $C = A \odot B$, noting that this operation does not depend on the choice of P, but does depend on 0 and I.

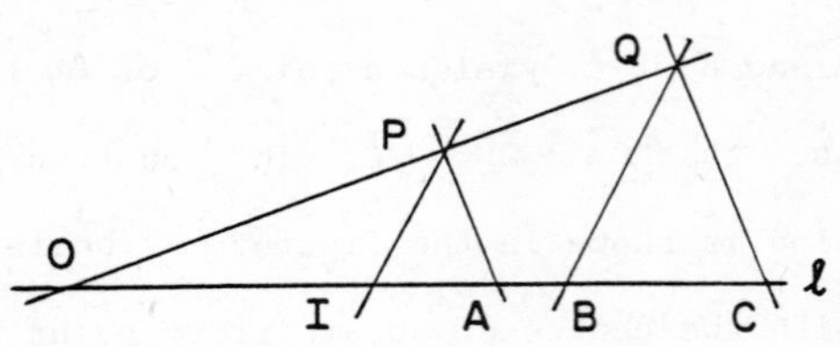

As before, we record a lemma.

LEMMA. Let $\alpha \in L$ leave the line ℓ invariant and fix the distinct points 0 and I of ℓ. Then, for all A, B on ℓ, $(A \odot B)\alpha = A\alpha \odot B\alpha$.

We now introduce a coordinate system.

LEMMA. Let $\alpha \in L$ leave the x-axis ℓ invariant and fix the points $0 = (0,0)$ and $I = (1,0)$. Then there is an automorphism γ of $\underline{R}$ such that, for all $x \in \underline{R}$, $(x,0)\alpha = (x\gamma,0)$.

Proof. Under the hypotheses, clearly $(x,0) \oplus (y,0) = (x + y,0)$, and since $|0I| = 1$, $(x,0) \odot (y,0) = (xy,0)$. Since the bijection α of E leaves that line ℓ invariant, it defines a bijection on ℓ, and the equation $(x,0)\alpha = (x\gamma,0)$ defines a bijection γ on $\underline{R}$. Now the two preceding lemmas tell us that $(x + y)\gamma = x\gamma + y\gamma$ and $(xy)\gamma = (x\gamma)(y\gamma)$, that is, that $\gamma \in \text{Aut } \underline{R}$. □

Since Aut $\underline{R} = 1$, we conclude that γ is the identity map on $\underline{R}$, that is, that γ fixes all points of ℓ. Discarding the coordinate system, which has served its purpose for the moment, we state the conclusion as follows.

THEOREM. If $\alpha \in L$ fixes two distinct points of a line ℓ, then it fixes all points of ℓ.

We can now complete the proof that $L \subseteq A$ and hence $A = L$. We choose a coordinate system in E. Let α be any element of L. Then some translation $\tau \in A$ carries $0 = (0,0)$ to $0\tau = 0\alpha$, whence $\alpha_1 = \alpha\tau^{-1}$ fixes 0. Next, some rotation $\sigma \in A$ with center 0 carries the x-axis ℓ to $\ell\sigma = \ell\alpha_1$, whence $\alpha_2 = \alpha_1\sigma^{-1}$ fixes 0 and maps ℓ to itself. Now $I\alpha_2$ lies on ℓ, whence some dilation $\delta \in A$ carries $I = (1,0)$ to $I\delta = I\alpha_2$, whence $\alpha_3 = \alpha_2\delta^{-1}$, mapping ℓ to itself, fixes both 0 and I. By the preceding lemma, α_3 fixes all points of ℓ.

Finally, since $J = (0,1)$ is not on ℓ, $J\alpha_3$ is not on ℓ. Now some $\beta \in A$ fixing all points of ℓ maps J to $J\beta = J\alpha_3$, whence $\alpha_4 = \alpha_3\beta^{-1}$ fixes all points of ℓ and also the point J. By the preceding lemma, since α_4 fixes 0 and J it fixes all points of the y-axis m. Let $P = (x,y)$ be any point of E. Since α_4 fixes the points $X = (x,0)$ on ℓ and $Y = (0,y)$ on m, it maps the line $m' = XP$ through X and parallel to m to itself, and the line $\ell' = YP$ through Y and parallel to ℓ to itself. Therefore it must map their intersection P to itself. We have shown that $\alpha_4 = 1$, that is, that $\alpha = \beta\delta\sigma\tau$, and $\alpha \in A$. □

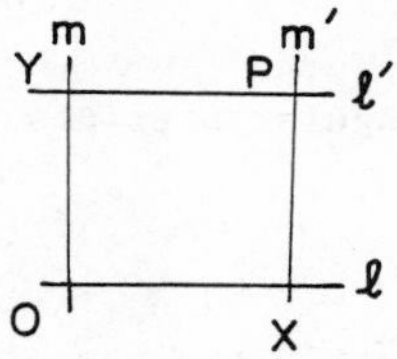

2. The coordinate plane over a field

We have not hesitated so far to rely on familiarity with the Euclidean plane E and its coordinatization by the field R of real numbers. We want now to formalize these ideas, for the analogous plane E(F) over an arbitrary field F. For the moment we regard E(F) as a set of points together with specification of certain sets of points called lines. (We could as well consider the set E(F) of points together with the ternary relation of collinearity, that three points lie on a common line.)

As expected, we define the set of points of E(F) to be the set of all ordered couples (x,y) for x, y ∈ F. The lines are then the loci

$\{(x,y) : ax + by + c = 0\}$ for all a, b, $c \in F$ such that not both a and b are 0.

The translation group $T(F)$ of E(F) consists of all transformations of E(F) of the form $(x,y) \mapsto (x + h, y + k)$ for arbitrary $h, k \in F$. Evidently $T(F) \simeq F^+ \oplus F^+$, where F^+ is the additive group of F. Let $A_0(F) = GL(2,F)$, the group of all linear bijections of E(F) as vector space with origin 0. Then we can define the affine group $A(F)$ of E(F) to be the group generated by $T(F)$ together with $A_0(F)$. Exactly as in the real case, $F = \underline{R}$, one sees that $A(F) = A_0(F)\cdot T(F)$, a semidirect product of $T(F)$ by $A_0(F)$, and that $A(F)$ is isomorphic to the group of all nonsingular matrices of the form $\begin{pmatrix} a_{11} & a_{12} & 0 \\ a_{21} & a_{22} & 0 \\ a_{31} & a_{32} & 1 \end{pmatrix}$ with entries from the field F.

As in the case that $F = \underline{R}$, we shall define $L(F)$ to be the group of all bijections of E(F) that map lines to lines. As we shall see, it is not always precisely true that $A(F) = L(F)$.

3. Incidence planes

We now abstract the combinatorial structure of a plane, based on the incidence relation J, that a point lies on a plane. We define an incidence plane to be a triple $G = (\Pi,\Lambda,J)$ where Π is a set of objects called points, Λ is a set of objects called lines, and J is a relation $J \subseteq \Pi \times \Lambda$; if $(P,\ell) \in J$ for some $P \in \Pi$ and $\ell \in \Lambda$, we say that the point P lies on the line ℓ.

Clearly the plane E(F) over any field F can be viewed as an incidence plane, and it is a remarkable fact that, conversely, any incidence plane that satisfies a few reasonable axioms is isomorphic to E(F) for some field F. We shall return to this below.

We begin by stating some simple axioms for an incidence plane G, which are obviously satisfied by any plane E(F).

Axiom A1. Given two distinct points there is exactly one line that contains them both.

COROLLARY. Two distinct lines have at most one point in common.

Proof. Suppose that lines ℓ_1 and ℓ_2 have in common two distinct points P_1 and P_2. Then, by Axiom A1, $\ell_1 = \ell_2$. □

Definition. Two lines will be called parallel if they are the same or if they contain no point in common.

Remarks. (1) There is a lack of symmetry between points and lines in that two distinct points always determine a unique line, while two distinct lines do not quite always determine a common point, indeed, only if they are not parallel. This lack of symmetry will be remedied when we pass to projective geometry.

(2) The definition of parallel lines is appropriate to the plane, but would be inappropriate in 3-dimensional space where there are 'skew lines' that do not meet but are not parallel in the sense of having the same direction. This latter possibility is excluded by the next axiom.

Axiom A2. Through any point there passes a unique line parallel to a given line.

COROLLARY. Parallelism is an equivalence relation on the set Λ of lines.

Proof. That the relation of parallelism is reflexive and symmetric is contained in the definition. To prove transitivity, suppose that three distinct lines are given, with ℓ_1 parallel to ℓ_2 and ℓ_2 parallel to ℓ_3. If ℓ_1 and ℓ_3 were not parallel, then, by the definition, they would contain a common point P, and we would have two distinct lines ℓ_1 and ℓ_3 through P parallel to ℓ_2, contrary to Axiom A2. □

Remarks. (1) As we have noted, Axiom A2 excludes the possibility that the 'plane' G have dimension n > 2, at least in the Euclidean case.

(2) Axioms A1 and A2 exclude the usual types of 'noneuclidean planes'. Thus A1 excludes 'spherical geometry' in which Π is the set of points of a sphere and Λ is the set of great circles, while A2 excludes the hyperbolic plane, which is discussed more fully in Chapter 9.

(3) Axioms A1 and A2 do not ensure that G has, in a reasonable sense, dimension as large as 2 -- in fact, they are satisfied if both Π and Λ are empty. The next axiom excludes this and several other degenerate cases, for example, exactly one point or exactly one line.

Axiom A3. There exist three points not all on the same line.

Example. We assume Axioms A1, A2, A3. Then there are three points A, B, C not on a common line. Then the three lines AB, BC, CA, given by Axiom A1, must be distinct. By Axiom A2, there is a line a through A parallel to BC and, using Axiom A2, we see that a cannot be any of AB, BC, CA.

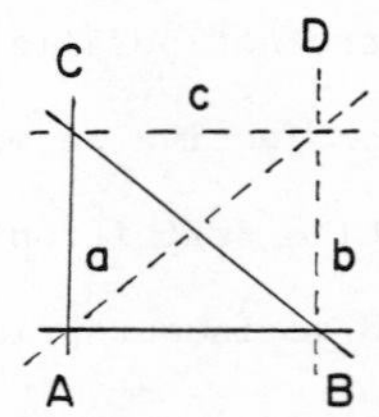

Similarly there must be a line b through B parallel to CA, and we see that this can be none of AB, BC, CA, a. Moreover, a and b cannot be parallel, whence they meet in a point D which must be distinct from A, B, C. Again, there must be a line c through C parallel to AB, and once more we see that this can be none of AB, BC, CA, a, b. But it is possible that AB, BC, CA, a, b, c are all the lines; in this case c also must contain D, whence a = AD, b = BD, c = CD, and there are only four points, A, B, C, D.

The easiest way to show that this finite incidence plane G satisfies the axioms is to note that here Π is a set of four elements, A, B, C, D, and Λ is the set of all 2-element subsets of Π. An interpretation more relevant to our purposes is to observe that G is isomorphic to the plane E(F) for F = $\underline{Z}_2$, the field of two elements 0 and 1. The four points are then (0,0), (1,0), (0,1) and (1,1), and the six lines fall into three pairs of parallels: x = 0 and x = 1; y = 0 and y = 1; x + y = 0 and x + y = 1.

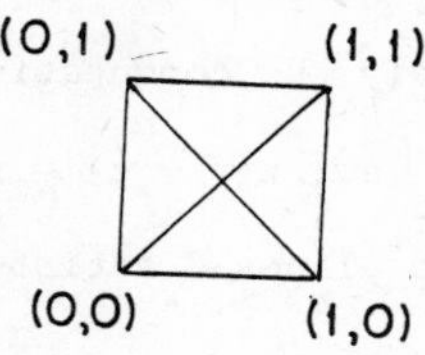

4. Introduction of coordinates

The Axioms A1, A2, A3 are all more or less obvious, and are commonly taken as the definition of an affine incidence plane, but they are not quite sufficient to prove that the incidence plane G is isomorphic to the plane E(F) for some field F.

The proof of this, under assumption of an addition Axiom 4 that we will state presently, consists mainly in constructing the field F, as a system of coordinates on some line ℓ in G. More specifically, we choose a line ℓ in G and two distinct points 0 and I on ℓ, and seek to define operations A $\oplus$ B and A $\odot$ B on ℓ by the constructions used earlier in the Euclidean case. These constructions can be copied literally to define such operations. The first obstacle is to prove that these two constructions yield operations independent of the choice of the point P used in the constructions: since this is true in the Euclidean case, we expect it to be true in the abstract case provided our axioms are strong enough, and it turns out that this is the case.

Our next objective is to show that the set F of all points on ℓ, equipped with the operations $\oplus$ and $\odot$, is a field. Once this is done, we can complete our proof that G is isomorphic to E(F) as in the Euclidean case. Again, most of the axioms for a field are routine and somewhat tiresome consequences of Axioms A1, A2, A3. However, it is noteworthy that certain of the axioms for a field do not follow from A1, A2, A3, notably the commutative law xy = yx for multiplication and the distributive law, x(y + z) = xy + xz. This can be shown by constructing incidence planes G satisfying A1, A2, A3 in which these axioms for the field F do not hold.

The striking fact, which we shall not prove, is that an incidence plane G, satisfying A1, A2, A3, is isomorphic to the plane E(F) for some field F if and only if G satisfies also a fourth Axiom A4, which is in fact the statement of the beautiful and classic <u>Theorem of Desargues</u>. We state this theorem in a form involving six points A, B, C, A', B', C', which we assume are distinct. We think of these as the vertices of two triangles ABC and A'B'C'. We state two conditions on these six points:

(C1) <u>the three lines</u> AA', BB', CC' <u>joining corresponding vertices of the two triangles either are parallel or all meet in a common point</u>.

(C2) <u>the three pairs</u> (AB,A'B'), (BC,B'C'), (CA,C'A') <u>of corresponding 'sides' of the two triangles either are three pairs of parallel lines or meet in three points that lie on a common line</u>.

Now Desargues' Theorem, which we take as <u>Axiom A4</u>, states that (C1) holds if and only if (C2) holds.

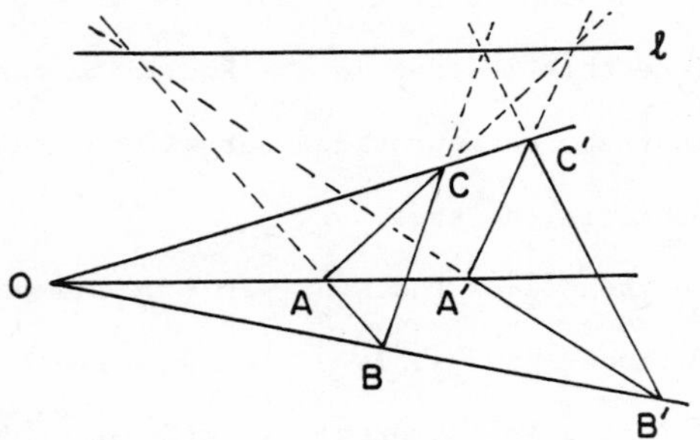

We now state the result cited above, whose proof we have only described in the roughest outline.

THEOREM. An incidence plane $G = (\Pi,\Lambda,J)$ is isomorphic to $E(F)$ for some (unique) field F if and only if G satisfies Axioms A1, A2, A3, A4.

We will in fact prove that A4 holds in the Euclidean plane E. However, both the statement and proof of Desargues' Theorem become simpler in the context of projective geometry, and we wait until we have discussed this subject in the next chapter before giving a proof.

5. The automorphism group of an incidence plane

We return to the question of the relation between the affine group $\mathcal{A}(F)$ for a field F and the group $\mathcal{L}$ of all bijections of E(F) that map lines to lines. If we regard E(F) as an incidence plane $G = (\Pi,\Lambda,J)$, which must then satisfy all four axioms, then $\mathcal{L}$ is simply its automorphism group Aut G.

As in the Euclidean case, it is trivial to see that $\mathcal{A}(F) \subseteq \text{Aut } G$. The argument in the other direction also goes through in the general case, up to a point. It shows that if $\alpha \in \mathcal{L}$ leaves the x-axis ℓ invariant and fixes the points $0 = (0,0)$ and $I = (1,0)$, then, for all $x \in F$, $(x,0)\alpha = (x\gamma,0)$ where γ is an automorphism of F. However, for general F, we cannot conclude that γ is the identity; for example, the complex numbers $\underline{C}$ admit the nontrivial automorphism of complex conjugation.

We nonetheless continue as in the Euclidean case, showing that $\alpha_4 = \alpha\theta$, for some $\theta \in A(F)$, fixes the point $J = (0,1)$ as well as 0 and I, and hence maps both the x-axis and y-axis to themselves. We now show that $(0,x)\alpha_4 = (0,x\gamma)$; this follows from the fact that since the line d through $X = (x,0)$ and $X' = (0,x)$ is parallel to IJ, so also is d_4 through $X\alpha_4$ and $X'\alpha_4$. Finally, given a point $P = (x,y)$, by dropping parallels through P to the two axes, we conclude the $P\alpha_4 = (x\gamma,y\gamma)$.

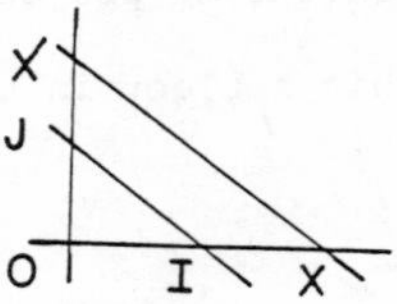

X'
J
O
I
X

To state our result, we define a subgroup $\mathrm{Aut}^* F$ of L to consist of all transformations $(x,y) \mapsto (x\gamma,y\gamma)$, for $\gamma \in \mathrm{Aut}\, F$; clearly $\mathrm{Aut}^* F$ is isomorphic to Aut F.

THEOREM. $L = \mathrm{Aut}\, G$ <u>is a semidirect product</u>, $\mathrm{Aut}\, G = \mathrm{Aut}^* F \cdot A(F)$ <u>of</u> $A(F)$ <u>by the group</u> $\mathrm{Aut}^* F \simeq \mathrm{Aut}\, F$.

COROLLARY. If Aut F = 1, then $L = \mathrm{Aut}\, G = A(F)$.

We have seen that Aut $\underline{R}$ = 1, and there are many other fields with no nontrivial automorphisms, for example, $\underline{Q}$ and all the fields $\underline{Z}_p$ for p a prime. On the other hand, we have noted that $\underline{C}$ does have a nontrivial automorphism. For much the same reason so does $\underline{Q}(\sqrt{2})$, the field of all numbers of the form $a + b\sqrt{2}$ for $a, b \in \underline{Q}$, and analogously the field $\underline{Z}_2(\omega)$ of the four elements of the form $a + b\omega$ for $a, b \in \underline{Z}_2$, where ω satisfies the equation $\omega^2 + \omega + 1 = 0$.

Problems.

Problem 1. State a theorem in elementary Euclidean geometry implying that the point $C = A \oplus B$ does not depend on the choice of the point P (not on ℓ) used in the construction. Give a 'proof' of this, assuming familiar facts about parallel lines, similar triangles, etc., but not using analytic geometry.

Problem 2. Show that not all three axioms A1, A2, A3 need to hold if E(F) is constructed as before but F is not a field, for example, if $F = \underline{Z}$ or if $F = \underline{Z}_4$, the integers modulo 4.

Problem 3. In what sense does Axiom A2 fail in spherical geometry? in hyperbolic geometry (as described in Chapter 3, Section 7)?

Problem 4. Describe the incidence planes that satisfy A1 and A2 but not A3.

Problem 5. (a) Let $F = \underline{Z}_3$. How many points are there in E(F), how many lines, how many points on a line, how many lines through a point, how many lines in a family of parallel lines? Make a schematic sketch of the points and lines of E(F).

(b) Let F be finite with m elements (m is necessarily a power of a prime). How many points and how many lines are there in E(F)?

Problem 6. If a field F_1 is contained in another field F_2, then, in a natural sense, $E(F_1)$ is contained in $E(F_2)$. What is the smallest subfield F of $\underline{R}$, containing the subfield $\underline{Q}$, with the following property? Let each of ℓ_1 and ℓ_2 be either a line in $E(\underline{R})$ through two points of

E(F) or a circle in E($\underline{R}$) with center in E(F) and through a point of E(F); then, if ℓ_1 and ℓ_2 meet in one or two points, these points are in E(F).

Problem 7. Let F = $\underline{Z}_2(\omega)$, the field of 4 elements described above. Find the orders of the groups T(F), A(F), A_0(F), and L(F).

References.

As will appear, the material in this chapter is more or less subsumed under projective geometry, the subject of the next. At this point we mention only two classics, the very readable book of Artin for the introduction of coordinates in an incidence plane, and Chapter 20 of M. Hall's Theory of Groups for a broader discussion of not necessarily desarguesian incidence planes.

CHAPTER SEVEN: PROJECTIVE GEOMETRY

1. Introduction

We begin with an informal discussion of the concept of a projection from one plane to another. Let E_1 and E_2 be two planes in 3-dimensional Euclidean space E^3 and 0 a point not on either plane. If P_1 is a point of E_1 and the line OP_1 from 0 through P_1 meets the plane P_2 in a point P_2 then we call P_2 the image of P_1 under (central) projection π from 0. Note that the line OP_1 may fail to meet E_2, in which case E_1 has no image under π, or the line OP_2, for P_2 in E_2, may fail to meet E_1, in which case P_2 is not the image of any point P_1 under π. (Both these things will happen unless E_1 and E_2 are parallel.) It is largely to remedy these defects in the map π that we introduce projective geometry, extending the planes E_1 and E_2 to projective planes E_1^* and E_2^* by adding new 'ideal' points in such a way that π extends naturally to a bijection π^* from E_1^* to E_2^*.

Suppose now that E_1 and E_2 are not parallel, hence meet in a line ℓ. The plane E_2' through 0 parallel to E_2 will meet E_1 in a line ℓ_1 parallel to ℓ, and it is precisely the points of ℓ_1 for which π is not defined. Similarly, the plane E_1' through 0 parallel to E_1 will meet E_2 in a line ℓ_2 parallel to

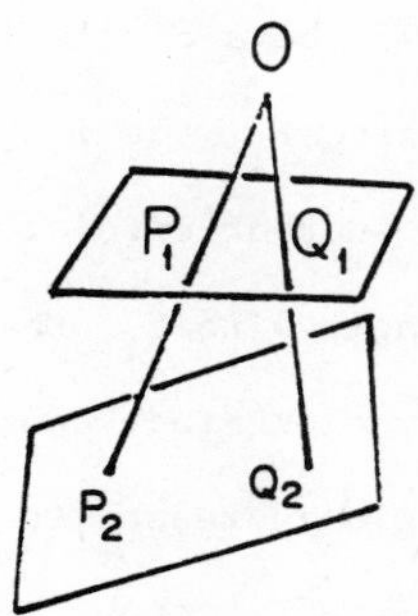

ℓ, and it is precisely the points of ℓ_2 which are not the images of any point of E_1 under π. In short, π is a bijection from $E_1 - \ell_1$ onto $E_2 - \ell_2$.

Let m_1 and n_1 be lines in E_1 meeting at a point P_1 on ℓ_1. Then, since P_1 has no image in E_2, the images m_2 and n_2 of m_1 and n_1 will have no point in common and will thus be parallel. When we postulate an image P_2 for P_1, it will thus be a common point of parallel lines. It will also be a 'point at infinity' in the sense that as a point Q of E_1 approaches P_1 along either m_1 or n_1, its image $Q\pi$ will recede indefinitely along m_2 or n_2.

The leading idea is to 'postulate' a 'line at infinity' ℓ_2^∞, adjoined to E_2, to serve as the image of ℓ_1 under π, and a 'line at infinity' ℓ_1^∞ adjoined to E_1, to serve as preimage of ℓ_2. Then π can be extended to a bijection from $E_1^* = E_1 \cup \ell_1^\infty$ to $E_2^* = E_2 \cup \ell_2^\infty$. We now abandon this somewhat awkward approach, which we hope has served its purpose as motivation, and give a more direct definition of a projective plane in the next section.

Before this, we want to mention the context in which these ideas arose classically, the study of 'conic sections'. Let O, E_1 and E_2 be as before, and let C_1 be a circle in E_1. If E_2 is parallel to E_1, the image C_2 of C_1 in E_2 will be another circle. If E_2 makes a small angle with E_1, then C_2 will be slightly elongated, and an ellipse. If this angle increases to the point where line OP_1, for some P_1 on C_1 is parallel to E_2, then P_1 will have no image in C_2, and C_2 will have become a parabola, a circle in which the image P_2 of P_1 has 'gone to

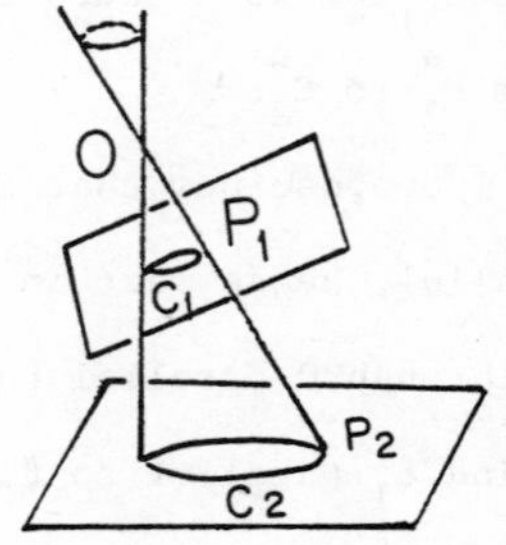

infinity'. As E_2 turns further, it will meet the infinite double cone at 0 with 'base' C_1 on both sides of 0, in two unconnected pieces, and C_2 will have become a hyperbola. This explains why conic sections of different types have many analogous properties.

2. Definition of the real projective plane

We saw above that a line ℓ through 0 meets the plane $E = E_1$ in a point P_1 unless ℓ is parallel to E_1, and we intend that the lines ℓ through 0 parallel to E_1 meet the extended plane E_1^* in a 'point at infinity'. We intend thus that the points of E_1^* be in bijective correspondence with the lines ℓ through 0. We seize the bull by the horns (as so often in mathematics) by defining E^* to be the set of all lines through 0 in E^3. More precisely, we define E^* to be an incidence plane $E^* = (\Pi,\Lambda,J)$ with Π, the set of 'points' of E^*, simply the set of lines through 0 in E^3. (In view of the 'points at infinity', the reader can easily believe that there is no hope of defining a reasonable metric on E^*.) The lines of E are the intersections of E with a plane through 0 not parallel to E, while the plane through 0 parallel to E is intended to correspond to the 'line at infinity' ℓ^∞. Thus we are led to define Λ to be the set of planes through 0. The definition of J is implicit in the above: an element of Π, that is a line ℓ through 0, is on (incident with) an element of Λ, that is a plane p through 0, just in case ℓ is contained in p.

To recapitulate, the real projective plane E^* is the incidence plane $E^* = (\Pi,\Lambda,J)$ where Π is the set of lines ℓ in E^3 through 0, Λ is the set of planes p in E^3 through 0, and $(\ell,p) \in J$, that is, ℓ is on p, just in case ℓ is contained in p.

<u>Remarks</u>. (1) It cannot be expected that E^* will satisfy all the axioms A1, 2, 3, 4 for the affine plane E, indeed, A2 will surely fail.

We now recast this definition in analytic form. We suppose a coordinate system given for E^3, with origin 0. A line ℓ through 0 is fully determined by any second point $P = (x,y,z) \neq (0,0,0)$ on ℓ, and two such points $P = (x,y,z)$ and $P' = (x',y',z')$ determine the same line ℓ if and only if they are proportional, that is, if $x' = kx$, $y' = ky$, $z' = kz$ for some $k \neq 0$. Accordingly we define an equivalence relation on the set of all triples $(x,y,z) \neq (0,0,0)$; $P \equiv P'$ if they are proportional in this sense, and take for Π the set of equivalence classes $[x,y,z]$ of triples $(x,y,z) \neq (0,0,0)$.

If p is a plane through 0 with equation $ax + by + cz = 0$ (where not all of a, b, c are 0), and P is on p, then any $P' \equiv P$ is also on p; thus we can define Λ to be the set of all loci $\{[x,y,z] : ax + by + cz = 0\}$ for $(a,b,c) \neq (0,0,0)$. This completes the analytic description of the projective plane E^*, as incidence plane.

The map η from E into E^* carrying (x,y) to $[x,y,1]$ is clearly injective; we may identify E with its image in E^*. If ℓ is a line in E with equation $ax + by + c = 0$, its image consists of all points of the line ℓ^* in E^* with equation $ax + by + cz = 0$, excepting the point $[b,-a,0]$, which can be viewed as the 'point at infinity' on ℓ. Now $E^* = E \cup \ell^\infty$, where ℓ^∞ is the 'line at infinity' with equation $z = 0$. This line ℓ^∞ consists of points $[1,m,0]$ as point at infinity lying on all the parallel lines $y = mx + k$ with slope m, together with the point $[0,1,0]$ lying on all vertical lines $x = h$.

To illustrate these ideas, if C is the locus in E of a polynomial equation $p(x,y) = 0$ and $p^*(x,y,z) = 0$ is obtained by substituting $x \mapsto \frac{x}{z}$, $y \mapsto \frac{y}{z}$ and clearing of fractions, the locus C^* in E^* of

$p^*(x,y,z) = 0$ consists of C together with its points at infinity. Thus the hyperbola $C_1 : x^2 - y^2 = 1$ goes to $C_1^* : x^2 - y^2 - z^2 = 0$, consisting of C_1 together for all of its points at infinity, obtained by taking $z = 0$, that is, the two points (1,1,0) and (1,-1,0), at infinity on the two asymptotes. The circle $C_2 : x^2 + y^2 = 1$ goes to $C_2^* : x^2 + y^2 - z^2 = 0$; this contains C_2, and, since $x^2 + y^2 + 0^2 = 0$ has only the excluded triple (0,0,0) as solution, it contains nothing more: the circle has no points at infinity. Note that C_1^* goes into C_2^* under an automorphism of E^*, in fact under interchange of x and z; in this sense the hyperbola and the circle are projectively equivalent.

3. The projective plane over an arbitrary field

Definition. If F is any field, the projective plane $E^*(F)$ over F is the incidence plane $E^*(F)$ defined as follows.

(1) Π is the set of equivalence classes $[x,y,z]$ on the set of all triples $(x,y,z) \neq (0,0,0)$, for $x, y, z \in F$, under the relation $(x,y,z) \equiv (x',y',z')$ if and only if $x' = kx$, $y' = ky$, $z' = kz$ for some $k \neq 0$ in F.

(2) Λ is a set in bijective correspondence with Π, whose elements we write as $\langle x,y,z \rangle$.

(3) $[x,y,z]$ is incident with $\langle a,b,c \rangle$ if and only if $ax + by + cz = 0$.

Example. Let $F = \underline{Z}_2$. Evidently $E^*(F)$ has exactly 7 points and 7 lines, while E(F) with 4 points and 6 lines is obtained from $E^*(F)$ by deleting the line ℓ^∞ and its 3 points (1,1,0), (1,0,0), and (0,1,0). In the (schematic) figure ℓ^∞ is shown as a broken line, and its three points as hollow points.

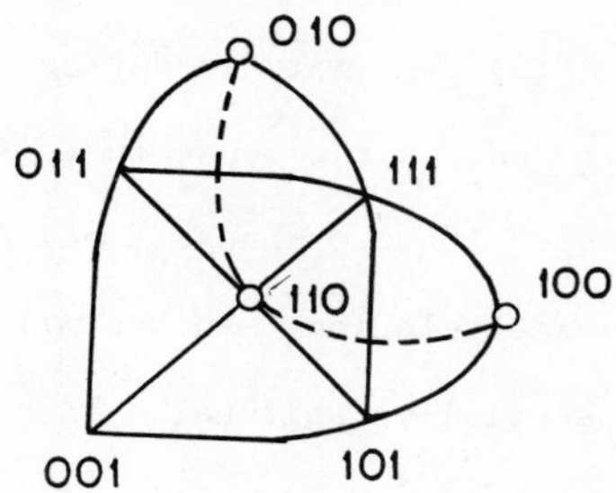

We now proceed, as with the affine planes, to characterize the projective planes $E^*(F)$ axiomatically. The following three axioms are analogous to those for the affine case, in fact, P1 is the same as A1.

Axiom P1. Given two distinct points there is exactly one line that contains them both.

Axiom P2. Given two distinct lines there is exactly one point that is contained in them both.

These are easily verified by inspection. For P1, if the two points lie in the affine subplane E(F), then a unique line of E(F), and so a unique line other than ℓ^∞, contains them both. If P_1 is in E(F) and P_2 on ℓ^∞, then the unique line is the line through P_1 with the slope (or direction) given by P_2. If both are on ℓ^∞, then the unique line is ℓ^∞. It is shorter, but less intuitive, to deduce P1 and P2 directly from the fact that, over any field, two independent linear homogeneous equations in three unknowns have a one dimensional subspace of solutions.

The next axiom excludes degenerate cases.

Axiom P3. There exist four distinct points no three of which lie on a common line.

Remarks. (1) It is easy to see that, using P1 and P2, Axiom P3 ensures the existence of the configuration shown, containing 7 points. If these 7 points are all, we have $E^*(\underline{Z}_2)$.

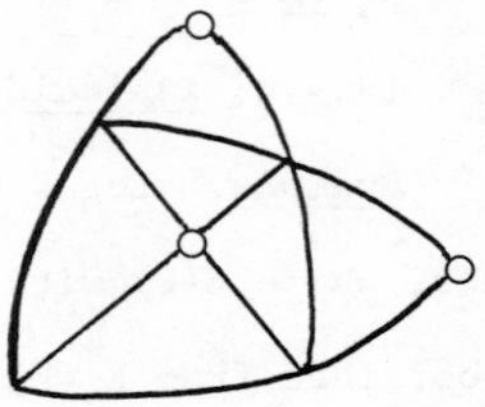

(2) The analytic definition of $E^*(F)$ is self dual in the sense that it remains unchanged if we exchange Π and Λ, replacing the incidence relation J by the inverse relation. In a similar sense, the set of Axioms P1, P2, P3, is (essentially) selfdual. Exchanging the words 'point' and 'line' interchanges P1 and P2, while it is easy to see that P3 is equivalent to its

dual P3': There exist four distinct lines, no three of which meet in a common point. It follows that if a certain theorem T is a consequence of these axioms, then so also is the dual assertion T', obtained by exchanging the words 'point' and 'line' in T.

In the case that F = R, we saw how to construct E^* from E by adjoining a line at infinity ℓ^∞ containing exactly one point on (the extension to E^* of) each line of a family of parallel lines of E. Conversely, we obtain E from E^* by deleting one line (and all points on it), the line ℓ^∞. The next theorem, which is easily verified by the same method, states the analogous result for axiomatically defined incidence planes.

THEOREM. Let E_1 be an affine incidence plane, satisfying axioms A1, A2, A3. Let E_1^* be constructed from E_1 by adjoining a new point on all members of each family of parallels in E_1, and one new line containing all these new points. Then E_1^* satisfies axioms P1, P2, P3. Conversely, if E_1^* is given satisfying P1, P2, P3, and E_1 is obtained by deleting one line from E^*, together with all points of that line, then E_1 satisfies A1, A2, A3.

In this sense the theories of affine planes and of projective planes are interchangeable, but the projective planes, although perhaps less familiar, are generally easier to study because of their simpler axioms and their higher degree of symmetry.

4. Coordinatization of projective planes

It follows from the above that a projective plane E_1^* is isomorphic to $E^*(F)$ for some field F if and only if the corresponding affine plane E_1 is isomorphic to E(F), thus if and only if the affine plane E_1 satisfies Desargues' Theorem. Now it is immediate that Desargues' Theorem for E_1 is equivalent to the projective form of Desargues' Theorem, which is simpler than the affine form, and, for a projective plane $E^*(F)$, easier to prove.

We state this theorem as an axiom.

Axiom P4. (Desargues' Theorem) Let A, B, C, A', B', C' be six distinct points, and let a, b, c be the (lines containing the) sides opposite A, B, C in the triangle ABC, and a', b', c' the sides opposite A', B', C' in the triangle A'B'C'. Then the following two conditions are equivalent:

(P4.1) the three lines AA', BB', CC' meet in a point;

(P4.2) the three points of intersection of a, a', of b, b', and of b, b' lie on a line.

From the discussion above we have the following.

THEOREM. An incidence plane E_1^* is isomorphic to $E^*(F)$ for some field F if and only if it satisfies the Axioms P1, P2, P3, P4.

In fact, we have proved the above only assuming the fact that E(F) satisfies the affine form of Desargues' Theorem, or, equivalently, that $E^*(F)$ satisfies P4. We now give a proof of this for $E^* = E(\underline{R})$.

The proof will illustrate the advantages of working in the projective plane. We first observe that the assertion that P4.1 implies P4.2 is dual to the converse implication, that P4.2 implies P4.1. Since $E^*(F)$ satisfies the principle of duality, if one of these implications holds, then so does the other. Thus it suffices to prove only one of the implications, and we choose to prove that P4.2 implies P4.1.

Since an affine plane isomorphic to E(F) can be obtained by deleting any chosen line of $E^*(F)$, we may suppose that the deleted line ℓ^∞ is the line mentioned in P4.2, that is, that the restrictions of a, a', of b, b', and of c, c' are three pairs of parallel lines in E(F). We must show then, in E(F), the following:

<u>If two triangles in the affine plane</u> E(F) <u>have their six vertices distinct, and corresponding sides parallel in pairs, then the three lines joining corresponding vertices either are parallel or meet in a point</u>.

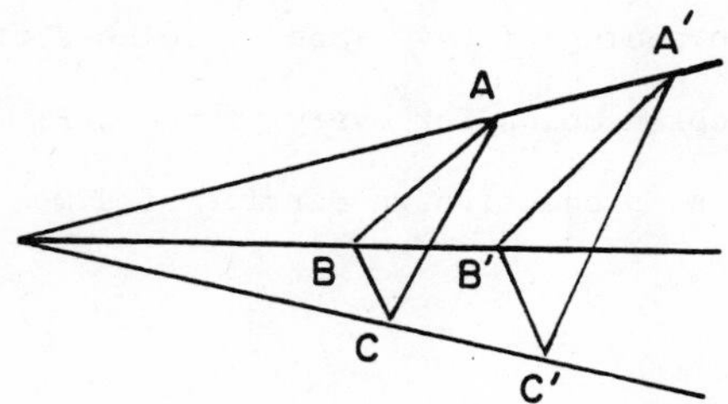

We now assume that F = <u>R</u>, where we can use the Euclidean metric and familiar properties of similar triangles; our proof could be translated into the affine geometry of E(F) for general F, although this could be rather laborious and not the simplest proof. To begin, we suppose that AA' and BB' are parallel, whence $|AB| = |A'B'|$, and the two similar triangles ABC and A'B'C' are in fact congruent. Now the sides BC and B'C' are parallel and equal, whence it follows that CC' also is parallel to BB'.

The case remains that no two of AA', BB', CC' are parallel. Let 0_{AB} be the intersection of AA' and BB'; then a suitable dilation δ_{AB} with center 0_{AB} carries the side AB to A'B', and hence the triangle ABC to the similar triangle A'B'C'. Likewise, a suitable dilation δ_{BC} with center the intersection 0_{BC} of BB' and CC' carries ABC to A'B'C'. But there is at most one dilation of the affine plane carrying a given triangle ABC to a second triangle A'B'C', whence $\delta_{AB} = \delta_{BC}$ and $0_{AB} = 0_{BC}$. This completes the proof of Desargues' Theorem for $E^*(\underline{R})$.

Before leaving this subject we make two remarks. First, there are many projective planes E_1^* satisfying P1, P2, and P3, but not P4, and these nondesarguesian planes have been studied in detail. Second, suppose the projective plane E_1^* occurs in a 3-dimensional projective space P. This is an incidence geometry with three kinds of objects, points, lines, and planes, and subject to a fairly obvious set of axioms analogous to the set P1, P2, P3, but no analog of P4. Then it follows from these axioms that Desargues' Theorem holds for every plane in P; in other words, there are no nondesarguesian projective geometries of dimension greater than 2.

5. The projective group

We now examine the automorphism group Aut $E^*(F)$ of the projective plane $E^*(F)$ over a field F. For this we recall our first definition of $E^*(F)$, given for the case $F = \underline{R}$, as the set of all lines through the origin 0 in the 3-dimensional coordinate space F^3, with the lines of $E^*(F)$ the planes through 0 in F^3.

We conclude that every automorphism of $E^*(F)$ is induced by an automorphism of F^3 that fixes 0, whence Aut $E^*(F)$ is a quotient group of the stabilizer $\text{Aut}_0\ F^3$ of 0 in the full group Aut F^3 of all automorphisms of F^3. Now the determination of these latter groups is entirely parallel to the case in dimension 2. The space F^3 is, by definition, the set of all points (x,y,z), for x, y, z $\in$ F. The planes of F^3 are the loci of equations $ax + by + cz + d = 0$, for a, b, c, d $\in$ F and not all of a, b, c = 0; the lines are the intersections of pairs of distinct nonparallel planes. An automorphism of F^3 is a bijection of F^3 that maps planes to planes, and so lines to lines. As in the 2-dimensional case, Aut F^3 is the split extension of a translation group $T(F^3)$ by $\text{Aut}_0\ F^3$. It is the group $\text{Aut}_0\ F^3$ that interests us, and, again as in the 2-dimensional case, this is a split extension of the general linear group GL(3,F) of dimension 3 over the field F by the group $\mathrm{Aut}^* F$ of all automorphisms of F of the form $(x,y,z) \mapsto (x\gamma, y\gamma, z\gamma)$ for $\gamma \in \mathrm{Aut}\ F$.

We thus have a homomorphism from $\mathrm{Aut}_0\ F^3 = \mathrm{Aut}^* F \cdot GL(3,F)$ onto $\mathrm{Aut}\ E^*(F)$, with kernel K the set of elements α of $\mathrm{Aut}_0\ F^3$ that leave all points of $E^*(F)$ fixed. But these are exactly the elements α which, for all (x,y,z) map (x,y,z) to some (kx,ky,kz), $k \neq 0$. Write $\alpha = \gamma\lambda$, $\gamma \in \mathrm{Aut}^* F$ and $\lambda \in GL(3,F)$. Since α maps the axes to themselves, so must γ. Since γ must fix $(0,0,0)$, $(1,0,0)$, $(0,1,0)$, $(0,0,1)$, $(1,1,1)$, it follows that λ must map these to multiples of themselves, hence λ must be a dilation $(x,y,z) \mapsto (kx,ky,kz)$ for some $k \neq 0$. Now these dilations, represented by scalar matrices $\begin{pmatrix} k & 0 & 0 \\ 0 & k & 0 \\ 0 & 0 & k \end{pmatrix}$, constitute precisely the <u>center</u> $ZGL(3,F)$ of $GL(3,F)$, consisting of all those elements that commute with all other elements.

We have seen that $ZGL(3,F) \subseteq K$, and that every element of K is of the form $\alpha = \gamma\lambda$ for $\gamma \in \mathrm{Aut}^* F$ and $\lambda \in ZGL(3,F)$. But this implies that $\gamma \in K$, that is, for all (x,y,z), one has $(x\gamma, y\gamma, z\gamma) = (kx,ky,kz)$ for some $k \neq 0$. Thus, for $x, y \neq 0$, $x\gamma/x = y\gamma/y$, and we conclude that, for some single $k \neq 0$, $x\gamma = k$ for all x. This is possible only if $k = 1$ and $\gamma = 1$. We have shown that $K = ZGL(3,F)$.

The standard notation is $PGL(3,F) = GL(3,F)/ZGL(3,F)$. We have shown the following.

THEOREM. <u>The automorphism group of the projective plane</u> $E^*(F)$ <u>over the field</u> F <u>is</u> $\mathrm{Aut}\ E^*(F) \simeq \mathrm{Aut}\ F \cdot PGL(3,F)$, <u>a split extension of the projective linear group</u> $PGL(3,F)$ <u>of dimension</u> 3 <u>over</u> F <u>by the group</u> Aut F <u>of all automorphisms of the field</u> F.

COROLLARY. <u>The automorphism group of the real projective plane</u> E^* <u>is isomorphic to</u> $PGL(3,\underline{R})$, <u>thus to the group of all nonsingular real 3-by-3 matrices modulo the subgroup of scalar matrices</u> aI, $a \neq 0$.

Notes. (1) A plane can intersect a double cone with apex 0 in the single point 0, in a single line through 0, or in a pair of intersecting lines at 0. If we permit 0 to 'go to infinity' the cone becomes a cylinder, and a plane can now fail to intersect or can intersect it in two parallel lines. These degenerate conic sections are the loci of equations $Ax^2 + Bxy + Cy^2 + Dx + Ey + F = 0$ for special values of the coefficients.

(2) Certain classical theorems are weaker consequences of Desargues' Theorem, and are equivalent to the validity of some, but not all, of the axioms for a field.

(3) A nondesarguesian affine plane, with associated nondesarguesian projective plane, is described in Problem 9.

Problems

Problem 1. Show that all nondegenerate conic sections in E are projectively equivalent in E^*.

Problem 2. Suppose that the field F has a finite number n of elements (n must then be a power of a prime). How many points and how many lines are there in $E^*(F)$? How many points on a line, how many lines through a point? What can be said if E_1 is merely an incidence plane satisfying P1, P2, P3 and there are only a finite number m of points on some line?

Problem 3. Discuss the degenerate cases excluded by Axiom P3.

Problem 4. Give a detailed proof of the theorem in Section 3.

Problem 5. Show that the automorphism group of the real projective plane E^* is transitive on points and, indeed, maps any three noncollinear points to any other three noncollinear points. What is the subgroup fixing all of three noncollinear points, and how does it act on the remaining points?

Problem 6. Show that the scalar matrices are the only 3-by-3 real matrices that commute with all other 3-by-3 real matrices.

Problem 7. We define the projective line $E^1(F)$ over a field F to consist of all equivalence classes [x,y] of pairs $(x,y) \neq (0,0)$ of elements $x, y \in F$ under the relation of proportionality. Let $F^* = F \cup \{\infty\}$, and map $E^1(F)$ to F^* by sending [x,y] to $\frac{x}{y}$ for $y \neq 0$ and [1,0] to ∞. Show that PGL(2,F) acts on F^* as follows: the coset M + ZGL(2,F), for $M = \begin{pmatrix} a & b \\ c & d \end{pmatrix}$ maps $x \in F^*$ to $\frac{ax + c}{bx + d}$ with the usual conventions regarding ∞. Now $E^1(F)$ or F^* has trivial incidence structure and no nontrivial invariant metric d(x,y). Show that, however, the cross ratio $CR(x_1,x_2,x_3,x_4)$ of four distinct elements, defined as $\frac{x_1 - x_3}{x_1 - x_4} \cdot \frac{x_2 - x_4}{x_2 - x_3}$ is invariant under PSL(2,F).

Problem 8. If $F = \underline{Z}_p$ or $F = \underline{Z}_2(\omega)$, $\omega^2 + \omega + 1 = 0$, the 4-element field, what is the order of Aut $E^*(F)$?

Problem 9. (See Albert and Sandler, Chapter 7, Section 3.) It is known that every affine plane with fewer than 9 points on each line satisfied Desargues' Theorem. A nondesarguesian affine plane E with 9 points on each lie can be constructed as follows. Let F be the plane with 9 elements, $F = \underline{Z}_3(\omega)$ where $\omega^2 = \omega + 1$. Let $P = \underline{Z}_3$, the 'prime subfield' $P \subset F$. We take the points of E' to be the same as those of E(F), $\Pi' = \Pi = \{(x,y) : x, y \in F\}$, but with a different set Λ' of lines.

Here $\Lambda' = \Lambda_1 \cup \Lambda_2$ consists of lines of two sorts. The lines ℓ in Λ_1 are all lines of E(F) with slope $m \notin P$, that is, all $\ell = \{(x,mx + b) : x \in F\}$ for $m, b \in F$ and $m \notin P$. The lines ℓ of Λ_2 are all sets $\ell = \{h + ms, k + mt) : s, t \in P\}$ for $h, k, m \in F$ and $m \neq 0$.

The main part of a verification that E_1 satisfies A1, A2, A3 consists in determining when two lines ℓ, ℓ' are parallel (that is, disjoint) and in verifying that otherwise they have exactly one point in common. We outline this verification. If ℓ, ℓ' are both in Λ_1, that is, are lines of E(F), then, as in E(F), they are parallel if and only if $m = m'$. Next suppose that both ℓ and ℓ' are in Λ_2. A common point then corresponds to a pair of solutions (s,s') and (t,t'), with $s, s', t, t' \in P$, of the equations (1) $h + ms = h' + m's'$ and (2) $k + mt = k' + m't'$. We rewrite each of these equations in the form $w = 0$, $w \in F$, and then take 'real and imaginary parts'; if $w = u + v\omega$, the equation $w = 0$ is equivalent to the two equations $u = 0$ and $v = 0$. One sees that each of these systems of two equations in two unknowns has a unique solution if $m \neq m'$, and at least one pair is incompatible if $mF = m'F$ unless $\ell = \ell'$. Thus ℓ and ℓ' are parallel if and only if $mF = m'F$ and otherwise have a single point in common. (Since ℓ depends only on mF, where $m \neq 0$, we could confine the choice of m to the set of values 1, ω, $1 + \omega$, $1 - \omega$, giving 4 families of parallel lines, with 9 lines in each family.

If $\ell \in \Lambda_1$ and $\ell' \in \Lambda_2$, a common point is given by a pair $s, t \in P$ satisfying $k' + m't = m(h' + m's) + b$, and a similar analysis shows, using the hypothesis $m \notin P$, that a unique solution always exists. In this case, ℓ and ℓ' always meet in a single point.

The first part of the problem is to supply full details of the argument above and to complete the verification that E_1 satisfies A1, A2, A3. The second part of the problem is to show that E_1 does not satisfy A4, Desargues' Theorem, or, alternatively, to show that E_1 is not isomorphic to E(F).

CHAPTER EIGHT: INVERSIVE GEOMETRY

1. Inversion in a circle

A Euclidean line may be regarded as the 'limiting case' of circles of increasing radius. In this sense, it is not surprising that reflection in a line is the analog of a transformation of inversion in a circle. In this section we give a metric definition of inversion in a circle and prove some of the basic properties of inversions; this seems to be the simplest approach, but, in the next section, we give a treatment more in the spirit of Euclidean geometry.

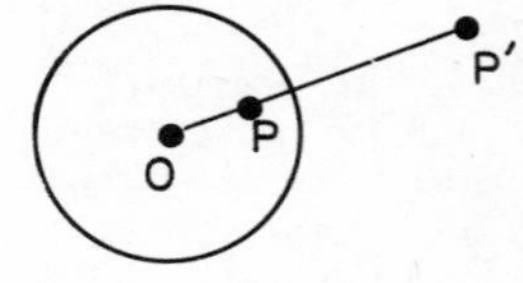

Let c be a circle in the Euclidean plane E with center 0 and radius r. If P is any point of E other than 0, we define P' to be the unique point on the line OP, on the same side of 0 as P is, such that $|OP| \cdot |OP'| = r^2$. Clearly the map $\gamma_c : P \mapsto P'$ is a bijection from E - {0} to E - {0}. As P approaches 0 along any path, that is, as $|OP|$ approaches 0, its image P' recedes indefinitely, that is, $|OP'|$ increases without bound. This leads us to define the <u>inversive plane</u> as $E^* = E \cup \{\infty\}$, the result of adjoining a new point ∞ to E, and to extend γ_c to a bijection from E^* to E^* by defining $0\gamma_c = \infty$ and $\infty\gamma_c = 0$.

Somewhat analogously to the case of the projective plane, we can construct E^* by projection, this time by <u>stereographic projection</u>. Let S be a sphere tangent to E at a point 0 and let 0' be the point of S opposite O. If P is any point of S other than O', the line O'P will meet E in a unique point P', the projection $P' = P\pi$ of P. Evidently π maps $S - \{O'\}$ bijectively onto E. If P approaches O' its image $P' = P\pi$ recedes 'to infinity', whence one is led to extend π to a bijection from S to E^* by setting $O'\pi = \infty$. (If hard pressed for a definition of E^*, one could use π to identify E^* with S.) It is natural to extend every line of E to include the additional point ∞.

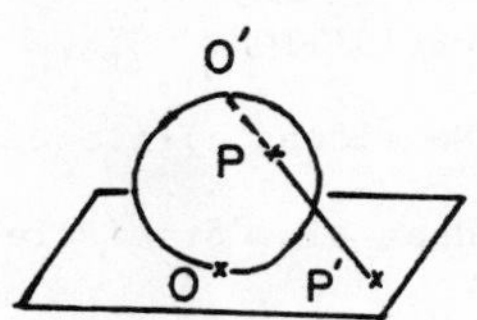

It is immediately clear from the definition that γ_c is an <u>involution</u>, that is, $\gamma_c^2 = 1$. Further, γ_c interchanges the interior and the exterior of c, while fixing every point of c.

THEOREM. <u>If ℓ is a Euclidean line or Euclidean circle, then $\ell\gamma_c$ is also a Euclidean line or Euclidean circle.</u>

<u>Proof</u>. We take O as the origin of a Cartesian coordinate system in E. Then the loci of equations $e : A(x^2 + y^2) + Bx + Cy + D = 0$, where not $AD = B = C = 0$, are precisely the Euclidean lines and circles. Now γ_c interchanges $O = (0,0)$ and ∞, and otherwise is given analytically by the transformations

$$x \mapsto \frac{x}{x^2 + y^2}, \qquad y \mapsto \frac{y}{x^2 + y^2}.$$

Substituting these values into e and multiplying through by a factor $x^2 + y^2$ yields an equation $e' : A + Bx + Cy + D(x^2 + y^2) = 0$, whose locus is again a Euclidean line or circle. □

COROLLARY. <u>Inversion γ_c in a circle</u> c <u>with center</u> O <u>maps every line through</u> O <u>to itself; it maps every line not through</u> O <u>to a circle through</u> O, <u>and every circle not through</u> O <u>to another circle not through</u> O.

<u>Proof</u>. This is immediate from the theorem together with the fact

that γ_c interchanges 0 and ∞. □

DEFINITION. Since inversion does not distinguish Euclidean lines from Euclidean circles, we use the term (inversive) line to mean either Euclidean line or circle, unless otherwise specified.

THEOREM. *If ℓ_1 and ℓ_2 are two lines meeting in a point $P \neq 0,\infty$, then the (directed) angle from $\ell_1\gamma_c$ to $\ell_2\gamma_c$ is equal but opposite to that from ℓ_1 to ℓ_2.*

Proof. Without loss of generality we may take c to be the unit circle in the complex plane $\underline{C} = E$. Then, for $z \neq 0$, $z\gamma_c = 1/\bar{z}$. Since the map $z \mapsto 1/z$ preserves angles (is conformal) while the conjugation map $z \mapsto \bar{z}$ reverses them, it follows that γ_c reverses angles. □

To avoid the exception, $P \neq 0,\infty$, in the theorem we decree, as is usual in analysis, that the angle between two (necessarily Euclidean) lines meeting at ∞ is the opposite of the angle between their images under γ_c at $\infty\gamma_c = 0$. □

2. Geometric treatment of inversion

We will now obtain an alternative definition of inversion in a circle and alternative proofs of the two theorems above by methods more in the spirit of Euclidean geometry. We begin by recalling two very elementary but not quite obvious lemmas from Euclidean geometry; in stating them, angles are taken as undirected, and the angle between two lines is taken as the smaller of the two angles between them.

LEMMA 1. *Let A, B_1, B_2 be three distinct points on a circle c with center O. Then $\angle B_1OB_2 = 2 \cdot \angle B_1AB_2$.*

Proof. Since the two triangles $[A,B_i,O]$, for $i = 1, 2$, are isosceles, we have $\angle AB_iO = \angle B_iAO = \alpha_i$, say, whence $\angle AB_iO = \pi - 2\alpha_i$.

Hence $\angle B_1OB_2 = 2\pi - (\angle AOB_1 + \angle AOB_2) =$
$2\alpha_1 + 2\alpha_2 = 2 \cdot \angle B_1AB_2$. □

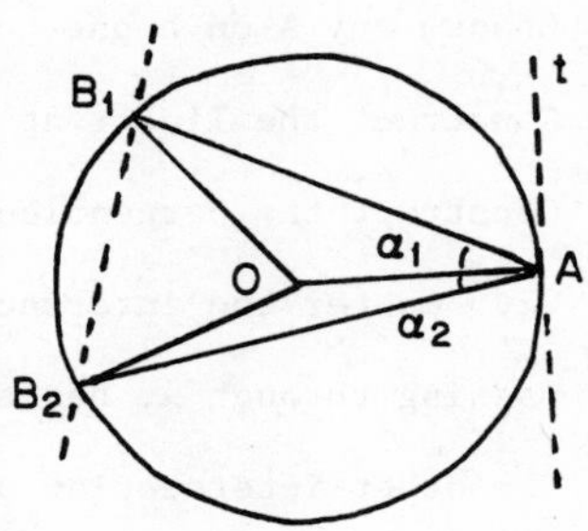

LEMMA 2. In the notation above, let t be the tangent to c at A. Then the angle between B_1A and t is $\angle AB_2B_1$.

Proof. As before, $\angle AOB_1 = \pi - 2\alpha_1$. By Lemma 1, $\angle AB_2B_1 = \frac{1}{2} \cdot \angle AOB_1 = \frac{\pi}{2} - \alpha_1$. Since OA is perpendicular to t at A, the angle from B_1A to t is $\frac{\pi}{2} - \angle B_1AO = \frac{\pi}{2} - \alpha_1$. □

THEOREM. Let a Euclidean line ℓ through the center O of a circle c meet a second circle c' in two points B_1 and B_2. Then c' is orthogonal to c if and only if $B_1\gamma_c = B_2$.

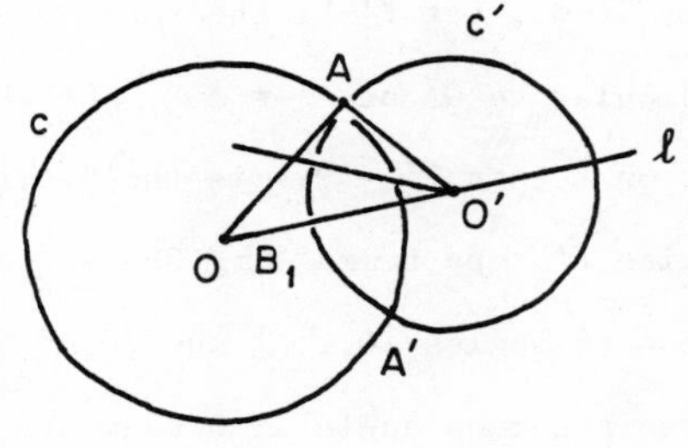

Proof. (1) Suppose first that c' with center O' is orthogonal to c, and meets c in A and A'. Let ℓ meet c' in B_1 and B_2. Since c and c' are orthogonal, OA is tangent to c' at A. By Lemma 2, $\angle OAB_2 = \angle OB_2A$. Since the triangles [O,A,B] and $[O,B_2,A]$ have also a common angle at O, they are similar, whence $\frac{|OB_1|}{|OA|} = \frac{|OA|}{|OB_2|}$, or $|OB_1| \cdot |OB_2| = |OA|^2 = r^2$.

(2) Suppose that $|OB_1| \cdot |OB_2| = r^2$, and now let OA be a line through O tangent to c' at some point A. As before, $|OB_1| \cdot |OB_2| = |OA|^2$, whence $|OA| = r$, A is on c, [O,A] is a radius of c and tangent to c', whence c and c' are orthogonal. □

Remark. This theorem provides an alternative definition of γ_c, and with it a simple construction, given B_1, for finding $B_2 = B_1\gamma_c$, as follows:

(1) Choose any A on c but not on OB_1;

(2) Construct the line t at A perpendicular to OA;

(3) Construct the perpendicular bisector m of B,A;

(4) With center the intersection O' of t and m construct a circle c' passing through A, hence also through B_1;

(5) The other intersection of c' sith OB_1 is now $B_2 = B_1\gamma_c$.

We now give a geometric proof of the theorem proved analytically above, that γ_c carries every (inversive) line to an (inversive) line.

<u>Proof</u>. (1) If the Euclidean line ℓ passes through O, then $\ell\gamma_c = \ell$.

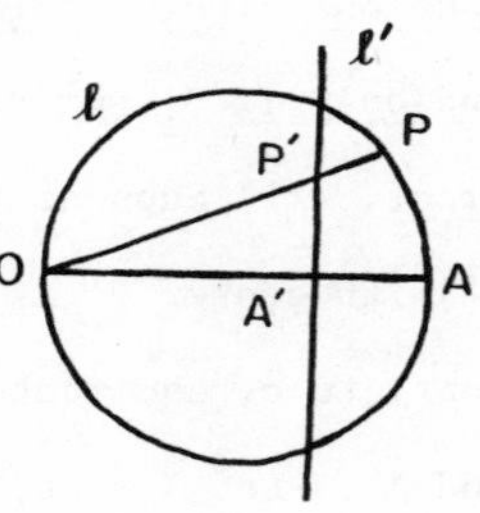

(2) Let ℓ be a circle through O and OA the diameter at O. Let ℓ' be the Euclidean line perpendicular to OA at $A' = A\gamma_c$. We show that, if P is on ℓ then $P' = P\gamma_c$ is the intersection of OP with ℓ'. By Lemma 1, $\angle OPA = \frac{\pi}{2} = \angle OA'P$. Since the triangles [O,P,A] and [O,A',P'] have also a common angle at O they are similar, whence $|OP| \cdot |OP'| = |OA| \cdot |OA'| = r^2$.

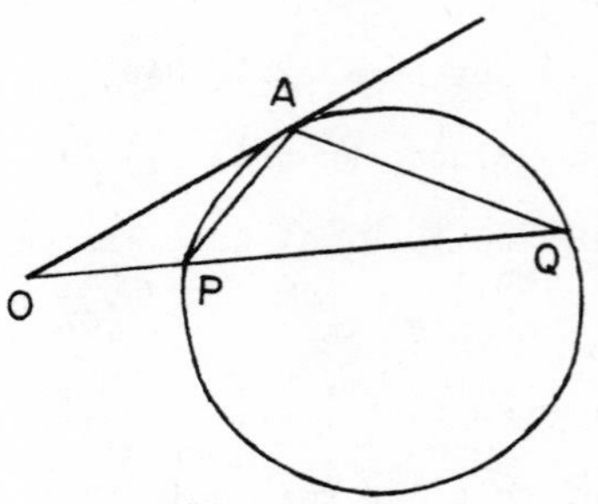

(3) Let ℓ be a circle not through O, and let OA be a tangent to ℓ through O, at some point A. Let a second line through O meet ℓ in distinct points P and Q. By Lemma 2, $\angle OAP = \angle OQA$; since the triangles [O,A,P] and [O,Q A] have also a common angle at O, they are similar, whence $|OP| \cdot |OQ| = OA^2$. Let δ be a dilation with center O and multiplier μ,

and let $P' = Q\delta$, hence $|OP'| = \mu \cdot |OQ|$. Then $|OP| \cdot |OP'| = \mu \cdot |OA|^2$. We now choose $\mu = \dfrac{r^2}{|OA|^2}$, whence $|OP| \cdot |OP'| = r^2$ and $P' = P\gamma_c$. This proves that $\ell\gamma_c$ is the circle $\ell\delta$. □

We give also a geometric proof of the other theorem proved analytically above, that γ_c reverses angles. First, we must give a geometric interpretation to the statement of the theorem.

A Euclidean circle ℓ is tangent, at a point P, to a line ℓ', either a Euclidean line or circle, if and only if ℓ and ℓ' have, in E^*, only the single point P in common. This gives a geometric definition of the tangent to a circle at a point. We now define the angle between two (inversive) lines ℓ_1 and ℓ_2 at a common point $P \neq 0, \infty$ by replacing either of ℓ_1, ℓ_2, if it is a circle, by its Euclidean tangent at P. With this, the theorem to be proved reduces to the case of two Euclidean lines ℓ_1 and ℓ_2 meeting at a point $P \neq 0, \infty$. If $\ell_0 = OP$, then $\angle \ell_1\ell_2$ is the sum or difference of $\angle \ell_0\ell_2$ and $\angle \ell_0\ell_1$, whence it suffices to treat the case that $\ell_2 = \ell_0$.

<u>Proof</u>. Let ℓ be a Euclidean line not passing through 0, and $P \neq 0$ a point on ℓ. Then $\ell\gamma_c$ is a circle meeting $\ell_0 = OP$ at 0 and $P\gamma_c$. Let t be the tangent to $\ell\gamma_c$ at 0 and t' that at $P\gamma_c$. Clearly OP makes opposite angles with $\ell\gamma_c$ at 0 and $P\gamma_c$, whence it remains only to show that OP makes equal angles with ℓ and t, that is, that ℓ and t are parallel. But $t = t\gamma_c$ and $\ell\gamma_c$, tangent at 0, have only the point 0 in common, whence t and ℓ have only the point $0\gamma_c = \infty$ in common, that is, t and ℓ are indeed parallel. □

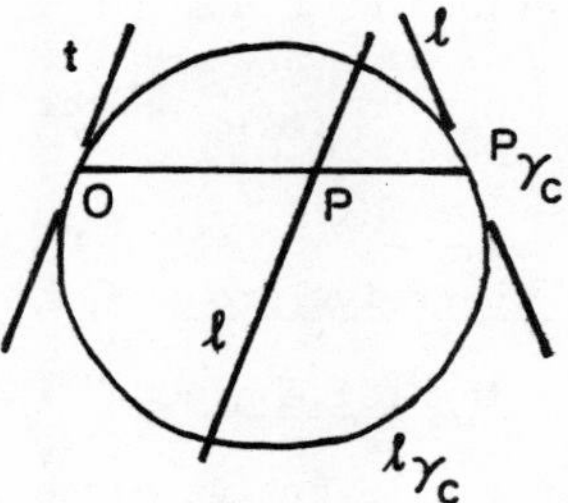

We conclude this section with a consequence of the above results. Before stating this, we observe that a uniform definition can be given for inversion γ_c in c, in the case that c is a circle, and reflection γ_c in c in the case that c is a Euclidean line.

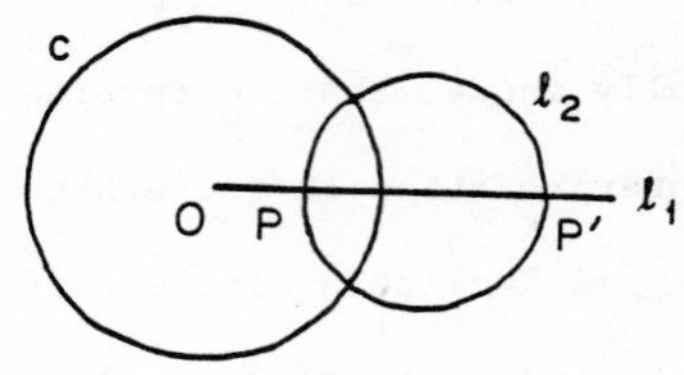

In fact, if c is any inversive line, then $P\gamma_c = P'$ if and only if P and P' are the two points of intersection of two inversive lines ℓ_1 and ℓ_2 both orthogonal to c.

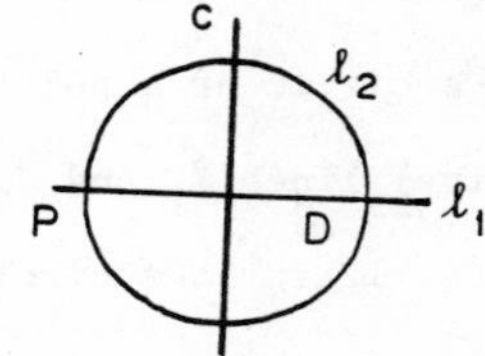

THEOREM. Let γ_c be inversion in an inversive line c, and let $\alpha \in M$. Then $\gamma_c^\alpha = \gamma_{(c\alpha)}$, inversion in the image $c\alpha$ of c under α.

Proof. Since M is generated by inversions, it suffices to treat the case that α is an inversion. Let $P' = P\gamma_c$. Then P and P' are the intersections of two inversive lines ℓ_1 and ℓ_2 orthogonal to c. Since α preserves angles, $P\alpha$ and $P'\alpha$ are the intersections of $\ell_1\alpha$ and $\ell_2\alpha$, orthogonal to $c\alpha$. Thus, for all P, $P\gamma_c\alpha = P'\alpha = P\alpha\gamma_{(c\alpha)}$, whence $\gamma_c\alpha = \alpha\gamma_{(c\alpha)}$ and $\gamma_c^\alpha = \gamma_{(c\alpha)}$. □

3. The inversive group

The inversive group M is defined to be the group of all transformations of the inversive plane E^* generated by all inversions in inversive lines. The subgroup M^+ of all orientation preserving elements of M is generated by all products of two inversions.

Our first goal is to obtain an analytic description of M and M^+. For this purpose we henceforth identify E with the complex plane $\underline{C}$, and thereby E^* with $\underline{C}^* = \underline{C} \cup \{\infty\}$. It will be shown that M^+ is exactly the group $LF(2,\underline{C})$ of all linear fractional transformations of the form

$$\alpha : z \mapsto \frac{az + c}{bz + d} \quad \text{for } a, b, c, d \in \underline{C} \text{ and } ad - bc \neq 0.$$

(It is understood that if $cz + d = 0$, then $z\alpha = \infty$, that $\infty\alpha = \frac{a}{b}$ if $b \neq 0$, and that $\infty\alpha = \infty$ if $b = 0$.)

The map ϕ carrying the matrix $M = \begin{pmatrix} a & b \\ c & d \end{pmatrix}$ to α is a homomorphism from $GL(2,\underline{C})$ onto $LF(2,\underline{C})$; it is to ensure this that we have interchanged the expected positions of b and c in the definition of α, in accordance with our convention that $\alpha\beta$ means α followed by β. Evidently α is trivial if and only if $b = c = 0$ and $a = d$, that is, if M is a scalar matrix $M = aI$ in the center $ZGL(2,\underline{C})$ of $GL(2,\underline{C})$. It follows that

$$LF(2,\underline{C}) \simeq PSL(2,\underline{C}) = GL(2,\underline{C})/ZGL(2,\underline{C}).$$

We note that if all of a, b, c, d are divided by a suitably chosen square root k of the determinant $ad - bc \neq 0$, then α is unchanged and now satisfies the condition that $ad - bc = 1$. Thus M is in $SL(2,\underline{C})$, and the kernel of the map ϕ restricted to $SL(2,\underline{C})$ is the center of $SL(2,\underline{C})$, consisting of the two matrices I and -I. Thus the representation of α by M in $SL(2,\underline{C})$ is almost unique, that is, unique up to change of sign. We have now

$$LF(2,\underline{C}) \simeq PSL(2,\underline{C}) = SL(2,\underline{C})/\{I,-I\}.$$

THEOREM. Let κ be complex conjugation, $\kappa : z \mapsto \bar{z}$. Then

$$\mathcal{M} \simeq \langle \kappa \rangle \cdot LF(2,\underline{C}), \text{ semidirect product,}$$

and
$$\mathcal{M}^+ \simeq LF(2,\underline{C}).$$

Proof. If α is given by $\begin{pmatrix} a & b \\ c & d \end{pmatrix}$, then α^κ is given by $\begin{pmatrix} \bar{a} & \bar{b} \\ \bar{c} & \bar{d} \end{pmatrix}$, whence $\langle LF(2,\underline{C}),\kappa \rangle = \langle \kappa \rangle \cdot LF(2,\underline{C})$, semidirect product.

We show first that $L = LF(2,\underline{C})$ is contained in $\mathcal{M}^+$. Since $\mathcal{M}$ contains all reflections γ_ℓ in Euclidean lines ℓ, $\mathcal{M}^+$ contains all products of two such reflections, hence all translations $z \mapsto z + B$, $B \in \underline{C}$, and all rotations $z \mapsto Az$, $A \in \underline{C}$ with $|A| = 1$. If γ_1 and γ_2 are inversions in two circles with center at 0 and radii r_1, r_2, then $\gamma_1\gamma_2 : z \mapsto (r_2/r_1)^2 z$ is a dilation, and it follows that M^+ contains all dilations $z \mapsto Az$, $A > 0$. It follows that M^+ contains all orientation preserving affine transformations of $\underline{C}$, that is all transformations $z \mapsto Az + B$, $A, B \in \underline{C}$, $A \neq 0$.

Since M contains inversion in the unit circle, $\gamma_{c_1} : z \mapsto 1/\bar{z}$, and inversion in the real axis, $\kappa : z \mapsto \bar{z}$, it contains their product $\eta : z \mapsto 1/z$. If, in α, $b = 0$, then α can be rewritten in the form $z \mapsto Az + B$, while, if $b \neq 0$, it follows that the transformation

$$z \mapsto -\frac{1}{b}\left(\frac{1}{bz + d}\right) + \frac{a}{b} = \frac{az + c}{bz + d}$$

is in M^+. Thus we have proved that $L \subseteq M^+$, whence $< L,\kappa > \subseteq M$.

We show next that $M \subseteq < L,\kappa >$, by showing that every inversion γ is in $< L,\kappa >$. We first note that, since $\eta : z \mapsto 1/z$ is in L, then $\gamma_{c_1} = \eta\kappa$, inversion in the unit circle, is in $< L,\kappa >$. Now let γ_c be inversion in any circle c. By some affine transformation $\alpha : z \mapsto Az + B$ in L, c_1 can be carried to c, whence $\gamma_c = \gamma_{c_1}^{\alpha}$, in $< L,\kappa >$. Second, let γ_ℓ be inversion in some Euclidean line ℓ. Now some affine α in L carries the real axis ℓ_1 to ℓ, whence $\gamma_\ell = \kappa^\alpha$, in $< L,\kappa >$. Thus $M \subseteq < L,\kappa >$.

Since $L \subseteq M^+ \subseteq M$ and $\kappa \in M$, we conclude that $< L,\kappa > \subseteq M$, whence $M = < L,\kappa >$. From $M = M^+ \cup M^+\kappa$ and $L \subseteq M^+$, which implies that $L \cap M^+\kappa = \emptyset$, we conclude that $L = M^+$. □

THEOREM. M^+ <u>is conformal</u>: <u>every</u> $\alpha \in M^+$ <u>preserves both the magnitude and sense of angles</u>: <u>every</u> α <u>in</u> M <u>but not in</u> M^+, <u>that is, every</u> $\alpha \in \kappa M^+$ <u>preserves the magnitude but reverses the sense of angles</u>.

<u>Proof</u>. This is immediate from the analytic representation of $\alpha \in M^+$ and the knowledge that analytic functions are conformal, but it also

follows geometrically from the earlier theorem that every inversion preserves the magnitude but reverses the sense of the angle between two inversive lines. □

THEOREM. <u>The stabilizer</u> M_∞ <u>of the point</u> ∞ <u>in</u> M <u>coincides in its action on</u> $\underline{C} = E$ <u>with the group</u> S <u>of similarities of the Euclidean plane</u> E, <u>whence</u> M_∞^+ <u>coincides with</u> S^+.

<u>Proof</u>. A transformation α in M fixes ∞ if and only if it is of the form $\alpha : z \mapsto Az + B$ or $\alpha : z \mapsto A\bar{z} + B$. Those of the first form constitute the group S^+, whence the group M_∞ of all such α of either form constitutes the group $S = S^+ \cup S^+\kappa$. □

THEOREM. M <u>is the group of all transformations of the inversive plane that map all inversive lines to inversive lines</u>.

<u>Proof</u>. We have shown that M is contained in the latter group G, and that $G_\infty = S = M_\infty$. For all P in $\underline{C}^*$ there exists an inversion γ_P in $G \cap M$ such that $P\gamma_P = \infty$. If γ is in G (or in M) and $\infty\gamma = P$, then $\gamma\gamma_P$ is in $G_\infty = M_\infty$, whence γ is in M (or in G). □

THEOREM. M^+ <u>is exactly triply transitive on the points of</u> $\underline{C}^*$.

<u>Proof</u>. It will suffice to show that M^+ is transitive on $\underline{C}^*$, that the stabilizer M_∞^+ of ∞ is transitive on $\underline{C}^* - \{\infty\} = \underline{C}$, and that the stabilizer $M_{\infty,0}^+ = M_\infty^+ \cap M_0^+$ of ∞ and 0 is exactly simply transitive on $\underline{C} - \{0\}$.

Since M^+ contains $\eta : z \mapsto 1/z$ carrying ∞ to 0 and $\gamma : z \mapsto z + B$ carrying 0 to all $B \in \underline{C}$, M^+ is transitive on $\underline{C}^*$. Since M_∞^+ contains all $\gamma : z \mapsto z + B$, M_∞^+ is transitive on $\underline{C}$. Now $M_{\infty,0}^+$ consists of all $\gamma : z \mapsto Az$, $A \neq 0$, hence for each $A \in \underline{C} - \{0\}$, a unique element γ carrying 1 to A, that is, is exactly transitive on $\underline{C} - \{0\}$. □

COROLLARY. <u>If</u> α <u>in</u> M^+ <u>fixes three distinct points, then</u> $\alpha = 1$.

<u>Proof</u>. Both α and the trivial map carry the three points P, Q, R to P, Q, R, whence $\alpha = 1$. □

COROLLARY. M^+ is transitive on the set of inversive lines.

Proof. Three distinct points P, Q, R of $\underline{C}^*$ determine a unique Euclidean line if one of them is ∞ or if all three are collinear; otherwise they determine a unique circle. Let ℓ and ℓ' be inversive lines, with P, Q, R distinct points on ℓ and P', Q', R' distinct points on ℓ'. Then α carrying P, Q, R to P', Q', R' must carry ℓ to ℓ'. □

Remark. M^+ is far from exactly transitive on inversive lines; this is equivalent to the fact that the stabilizer M^+_ℓ in M^+ of an inversive line ℓ is far from trivial. (Indeed, the study of these groups M^+_ℓ is the subject of the two remaining chapters.) This is not surprising, since we have seen that the stabilizer M^+_P of a point in M^+ is isomorphic to the Euclidean similarity group S, and the study of this group embraces the geometry of the Euclidean plane.

4. Classification of elements of M^+

We seek a classification of the elements of M^+ that is 'geometric' in the sense that it is invariant under 'change of coordinates', that is, under conjugation within M^+. As in the Euclidean case, this classification is based on the set of fixed points of the transformation.

The trivial transformation 1 is in a class by itself, since it fixes all points of $\underline{C}^*$.

The fixed points of α are the roots of the equation $z = \dfrac{az + c}{bz + d}$ where $ad - bc = 1$. Since $bz + d = 0$ implies that $z = \infty \neq z$, we may multiply by $bz + d$ to replace this equation by $bz^2 - (a - d)z - c = 0$. If $b \neq 0$, this equation has two distinct roots if $(a - d)^2 + 4bc = (a + d)^2 - 4ad + 4bc = (a + d)^2 - 4 \neq 0$, and has only one root if $(a + d)^2 = 4$. If $b = 0$, then ∞ is a fixed point as is $\dfrac{c}{d - a}$. These are distinct unless $d = a$, whence from $ad = 1$, $a = d = \pm 1$ and $(a + d)^2 = 4$.

THEOREM. <u>If $\alpha \neq 1$, then α is parabolic, with a single fixed point, if</u> $(a+d)^2 = 4$, <u>and otherwise is loxodromic, with two fixed points.</u>

We seek a canonical form for parabolic α under conjugation. Let parabolic α have fixed point P, and let γ_1 in M^+ map P to ∞. Then $\alpha_1 = \alpha^{\gamma_1}$ has fixed point ∞, hence is of the form $z \mapsto Az + B$. Now this has a second fixed point if $A \neq 1$, whence we conclude that α_1 has the form $z \mapsto z + B$. Now conjugation by $\gamma_2 : z \mapsto B^{-1}z$ carries α_1 to $\alpha_2 = \alpha_1^{\gamma_2} : z \mapsto z + 1$.

THEOREM. <u>All parabolic elements are conjugate to the transformation</u> $z \mapsto z + 1$.

Now let α be a loxodromic transformation. After conjugation we may suppose that α has fixed points 0 and ∞, hence has the form $\alpha : z \mapsto Az$. Evidently $b = c = 0$, whence $ad = 1$, and $A = a/d = a^2$. We may write $a = re^{i\theta}$, $r > 0$, whence $A = r^2e^{2i\theta}$. In the special case that $\theta = 0$, α is a dilation with center 0 and multiplier $A = r^2$. In the special case that $r = 1$, α is a rotation $z \mapsto e^{2i\theta}z$ about 0 through an angle 2θ; we note that α will have finite order if and only if θ is a rational multiple of π. In the general case, where neither $\theta = 0$ nor $r = 1$, the orbits $\{P\alpha^n : n \in \underline{Z}\}$ lie on spirals $z = r^te^{i\theta(t-t_0)}$ winding from 0 to ∞; this accounts for the term 'loxodromic'.

Since the two eigenvalues $re^{\pm i\theta}$ of the matrix for loxodromic α are unchanged by conjugation, and A is the quotient of one by the other, A is invariant under conjugation apart from possibly replacing A by A^{-1}, that is, α by α^{-1}. Further, since the square t^2 of the <u>trace</u> $t = a + d$ is invariant under conjugation (the characteristic polynomial is $x^2 - tx + 1$), and, in the canonical form above $ad = 1$, $t^2 = A + 2 + A^{-1}$, the value of t^2 determines the pair A, A^{-1} uniquely.

THEOREM. <u>Two loxodromic elements are conjugate if and only if they have the same value for</u> $(a + d)^2$. <u>Each loxodromic element is conjugate to</u>

<u>a unique pair of elements of the forms</u> $z \mapsto Az$, $z \mapsto A^{-1}z$ <u>where</u> $A \neq 0, 1$.

In the figures below we describe typical elements α of M^+ by showing the <u>flow lines</u> of α; these are lines invariant under α. The ends of a flow line are fixed points, and α moves all points of the line (except its end points) along the line in the same direction. If the flow line is a circle without fixed points, all points are moved around the circle in the same direction. We first show a parabolic element α with fixed point P, first for $P \in \underline{C}$ and second for $P = \infty$.

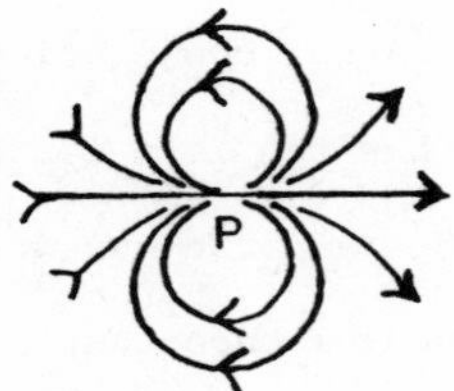

Parabolic, $P \in \underline{C}$

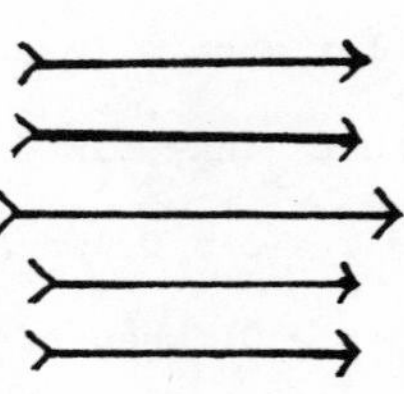

Parabolic, $P = \infty$

We next show two special types of loxodromic transformation that arise if the coefficients a, b, c, d are all real. Then α is <u>elliptic</u> if $t^2 < 4$ and <u>hyperbolic</u> if $t^2 > 4$. The next two figures show elliptic α with fixed point P, $Q \in \underline{C}$ and with P, $Q = 0, \infty$.

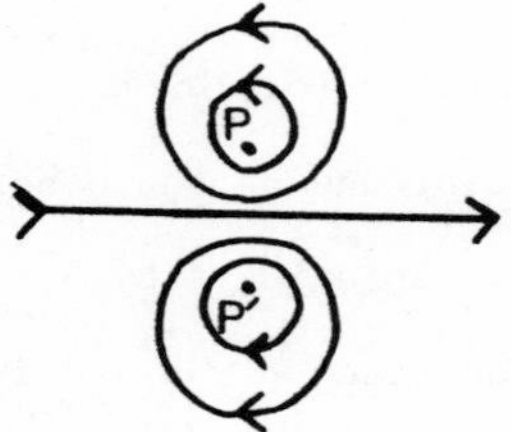

Elliptic, P, $Q \in \underline{C}$

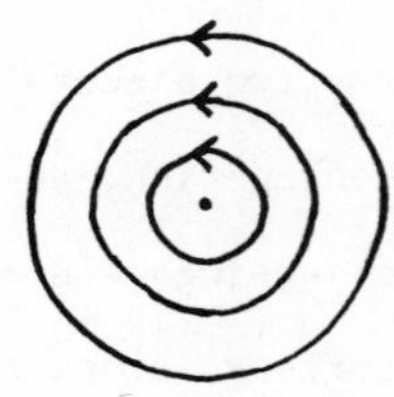

Elliptic, P, $Q = 0, \infty$

The next two figures show hyperbolic α with fixed points P,Q.

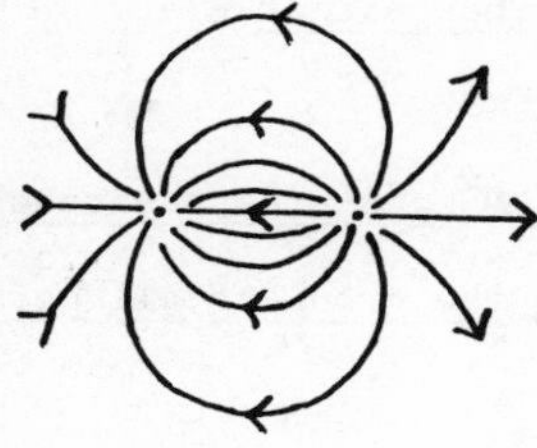

Hyperbolic, P, Q $\in \underline{C}$

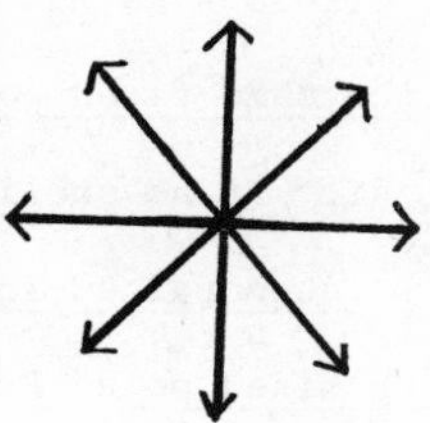

Hyperbolic, P, Q = 0, ∞

Finally, the general loxodromic transformation α is in a certain sense a blend of the elliptic and hyperbolic types, with flow lines that are spirals. We show such α with fixed points P and Q.

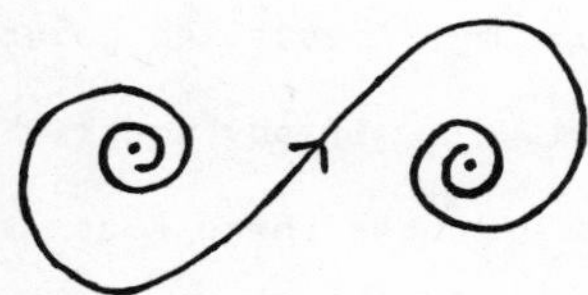

Loxodromic, P, Q $\in \underline{C}$

Loxodromic, P, Q = 0, ∞

5. Solvable subgroups of M^+

We make repeated use of the following.

LEMMA. _If a permutation_ α _of a set has fixed point set_ F, _and_ β _is a second permutation of the same set, then_ α^β _has fixed point set_ $F\beta$.

Proof. If α fixes P, that is, $P\alpha = P$, then $(P\beta)(\alpha^\beta) = P\alpha\beta = P\beta$, and α^β fixes $P\beta$. □

COROLLARY. _If_ $\alpha\beta = \beta\alpha$, _then_ β _permutes the fixed point set of_ α.

Suppose now that α has a single fixed point P and that $\alpha\beta = \beta\alpha$. Then β must fix P. If β had exactly two distinct fixed points P and Q, then α, fixing P and permuting the set {P,Q}, would have to fix Q, contrary to hypothesis. This proves the following.

LEMMA. <u>If</u> $\alpha\beta = \beta\alpha$ <u>and</u> α <u>is parabolic, then either</u> $\beta = 1$ <u>or</u> β <u>is parabolic with the same fixed point</u> P.

COROLLARY. <u>If an abelian subgroup</u> G <u>of</u> M^+ <u>contains a parabolic element, then</u> G <u>is contained in</u> T_P, <u>for some</u> P, <u>the group of all parabolic elements with fixed point</u> P <u>together with</u> 1.

Suppose now that α has exactly two fixed points P and Q. If $\alpha\beta = \beta\alpha$ then β must fix the set $\{P,Q\}$. If β does not fix both P and Q, then it must interchange them. After conjugation we may suppose that $P = 0$, $Q = \infty$, whence α has the form $\alpha : z \mapsto Az$ for some $A \in \underline{C}$, $A \neq 0, 1$, and that β is the transformation $\beta : z \mapsto \frac{1}{z}$. Now $\alpha^\beta = \alpha^{-1}$, and $\alpha\beta = \beta\alpha$ requires that $\alpha^2 = 1$, hence $A = -1$. Now α and β generate a four group $V = \{1,\alpha,\beta,\gamma\}$ where $\alpha : z \mapsto -z$ has fixed points $0, \infty$; $\beta : z \mapsto \frac{1}{z}$ has fixed points $+1$, -1; and $\gamma = \alpha\beta : z \mapsto -\frac{1}{z}$ has fixed points $+i, -i$. If any nontrivial δ commutes with α and β, it must preserve each of these three sets of fixed points. Since it cannot fix all six points, we may suppose that it interchanges 0 and ∞; then δ has the form $\delta : z \mapsto \frac{B}{z}$ for some $B \neq 0$, and $\beta\delta = \delta\beta$ implies that $B = \pm 1$, whence $\delta = \beta$ or $\delta = \gamma$. We have shown the following.

LEMMA. <u>If an abelian subgroup</u> G <u>of</u> M^+ <u>contains a loxodromic element with fixed points</u> P <u>and</u> Q, <u>then</u> G <u>is contained either in</u> $M^+_{P,Q}$ <u>or in a conjugate of the four group</u> V <u>described above</u>.

The <u>normalizer</u> N(G) of a subgroup G of M^+ is the group of all elements δ in M^+ such that $G^\delta = G$.

Suppose that $1 \neq G \subseteq T_P$. If $1 \neq \alpha \in T_P$ and $\alpha^\beta \in G$ then both α and α^β are parabolic with fixed point P, whence β fixes P, that is, $N(G) \subseteq M^+_P$. In particular, $N(T_P) = M^+_P$. If $G \subseteq M^+_P$ and $G \cap T_P \neq 1$, the same argument shows that $N(G) \subseteq M^+_P$, whence in particular, $N(M^+_P) = M^+_P$.

Suppose that $1 \neq G \subseteq M^+_{\{P,Q\}}$. Then $N(G) \subseteq M^+_{\{P,Q\}}$. In particular, $N(M^+_{P,Q}) = N(M^+_{\{P,Q\}}) = M^+_{\{P,Q\}}$.

Finally, suppose that $\delta \in N(V)$, that is, $V^\delta = V$. Then δ must permute the three sets $F_\alpha = \{0,\infty\}$, $F_\beta = \{+1,-1\}$, $F_\gamma = \{+i,-i\}$. Now the transformation $z \mapsto \frac{i}{\sqrt{2}}\frac{z-1}{z+1}$ permutes F_α, F_β, and F_γ cyclically, while the transformation $z \mapsto \frac{1}{\sqrt{2}}\frac{z-1}{z+1}$ interchanges F_α and F_β, while fixing +i and -1. These two transformations generate a group S effecting by conjugation the full automorphism group of V, where Aut V $\simeq$ Sym $\{F_\alpha, F_\beta, F_\gamma\}$, the symmetric group on three objects. We conclude that $N(V) = S\cdot V$, semidirect product of V by S and that N(V) is its own normalizer. (We note that N(V) has order $6\cdot 4 = 24$; it is in fact isomorphic to the alternating group A_4, or to $\mathrm{Sym}^+ \square_3$, viewed as permuting the 4 diagonals of the cube $\square_3$.)

THEOREM. Let $1 \neq G_1 \vartriangle G_2 \vartriangle \cdots \vartriangle G_t$ be a chain of subgroups of M^+, where G_1 is abelian and, for $1 \leqslant i < n$, each G_i is a normal subgroup of G_{i+1}. Then G_t is contained in some M^+_P, some $M^+_{\{P,Q\}}$, or in some conjugate of N(V).

Proof. If $G_1 \subseteq T_P$, then, inductively, all $G_{i+1} \subseteq N(G_i) \subseteq M^+_P$. If $G_1 = V$, then, inductively, all $G_{i+1} \subseteq N(V)$. The remaining possibility is that $G_1 \subseteq M^+_{P,Q}$ with no $G_i = V$, whence, inductively, all $G_{i+1} \subseteq M^+_{P,Q}$.

A group G_t is solvable if it has a chain of subgroups as above, with the additional condition that all quotients G_{i+1}/G_i are abelian.

COROLLARY. The solvable subgroups of M^+ are exactly the subgroups of groups M^+_P, groups $M^+_{\{P,Q\}}$, and conjugates of N(V).

Proof. It remains to show that these groups are solvable. Taking $P = \infty$ one sees that $M^+_P/T_P \simeq \underline{C}^\times$, the multiplicative group of nonzero complex numbers. The quotient group $M^+_{\{P,Q\}}/M^+_{P,Q}$ is isomorphic to the symmetric group on the set {P,Q}, hence of order 2. The quotient $N(V)/V \simeq S$, and S is not abelian; however S contains a normal subgroup A of order 3 with S/A of order 2. We have a series $G_1 = V$, $G_2 = A \cdot V$, $G_3 = N(V)$, with $G_2/G_1 \simeq A \simeq C_3$ and $G_3/G_2 \simeq S/A \simeq C_2$. □

6. Finite subgroups of M^+

If G is a finite subgroup of M^+, then every nontrivial element of G has finite order, hence is elliptic, with two fixed points P and Q. Moreover, the stabilizer $G_{P,Q} = G_P \cap G_Q$ is a finite cyclic group of order $p \geq 2$. Similarly, if G is a finite group of rotations of the sphere S, then every nontrivial element of G has two fixed points P and Q, and $G_{P,Q}$ is cyclic of finite order.

We shall enumerate all finite groups G of permutations of a set Ω such that G is finite, and each nontrivial element of G has two fixed points P and Q, with $G_{P,Q}$ cyclic. We note that this abstract description applies equally to the finite groups of rotations of the sphere.

Let $\Phi \subseteq \Omega \times \Omega$ be the set of all 'axes' $\phi = F(\alpha) = \{P,Q\}$, for $1 \neq \alpha \in G$. Since $F(\alpha)\beta = F(\alpha^\beta)$, G permutes the set Φ. For $\phi \in \Phi$, let G_ϕ be the stabilizer of ϕ under this action, and let $[\phi] = \phi G = \{\phi\alpha : \alpha \in G\}$, the orbit of ϕ under G. Under the map from G to $[\phi]$ carrying α to $\phi\alpha$, the inverse image of $\phi\alpha$ is the coset $G_\phi\alpha$, of cardinal $|G_\phi|$. Therefore $|[\phi]| = |G|/|G_\phi|$.

If $\phi = \{P,Q\}$, let the cyclic group $G_{P,Q}$ have order $P_\phi \geq 2$. Then either $G_\phi = G_{P,Q}$ or $G_\phi = G_{P,Q} \cup G_{P,Q}\alpha$, where α transposes P and Q. Thus $|G_\phi| = d_\phi p_\phi$ where $d_\phi = 1$ or $d_\phi = 2$.

Now $G - 1 = \bigcup_{\{P,Q\} \in \Phi} (G_{P,Q} - 1)$, disjoint union, whence

$$|G| - 1 = \sum_{\phi \in \Phi} (p_\phi - 1) = \sum_\phi |[\phi]| \cdot (p_\phi - 1) = \sum_\phi \frac{|G|}{|G_\phi|}(p_\phi - 1),$$

and

$$1 - \frac{1}{|G|} = \sum_{\phi} t_{\phi} \qquad \text{where} \qquad t_{\phi} = \frac{p_{\phi} - 1}{d_{\phi} p_{\phi}} \quad .$$

Evidently $t_{\phi} \geqslant \frac{2-1}{2 \cdot 2} = \frac{1}{4}$, whence, from $\Sigma\, t_{\phi} = 1 - \frac{1}{|G|} < 1$, we conclude that the number n of classes $[\phi]$ is at most 3.

We write G_i, p_i, d_i, t_i instead of G_{ϕ_i}, p_{ϕ_i}, d_{ϕ_i}, t_{ϕ_i}, $1 \leqslant i \leqslant n$.

Case 1. If $n = 0$, then $G = 1$. If $n = 1$, then $G = G_1$, a cyclic group.

Next suppose that $n = 2$. If $d_1 = d_2 = 1$, then $t_1, t_2 \geqslant \frac{1}{2}$, contrary to $t_1 + t_2 < 1$. If $d_1 = d_2 = 2$, then $t_1 + t_2 = 1 - \frac{1}{2p} - \frac{1}{2q}$, whence $\frac{1}{|G|} = \frac{1}{2p_1} + \frac{1}{2p_1}$. But $|G| > |G_1| = 2p_1$ whence $\frac{1}{|G|} < \frac{1}{2p_1}$, a contradiction. Thus we may suppose that $d_1 = 1$ and $d_2 = 2$. Since $t_2 \geqslant \frac{1}{4}$, $t_1 < \frac{3}{4}$, which implies that $p_1 < 4$, that is, $p_1 = 2$ or $p_1 = 3$.

Case 2. $p_1 = 2$, $p_2 = 2$, $t_1 = \frac{1}{2}$, $t_2 = \frac{p_2 - 1}{2p_2}$. Now $t_1 + t_2 = 1 - \frac{1}{|G|}$ gives $|G| = 2p_2 = |G_2|$, whence $G = G_2$, a dihedral group of order $2p_2$.

Case 3. $p_1 = 3$, $t_1 = \frac{2}{3}$, $t_2 = \frac{p_2 - 1}{2p_2}$. Now $t_1 + t_2 < 1$ implies that $p_2 < 3$, whence $p_2 = 2$, $t_2 = \frac{1}{4}$, and $|G| = 12$. There are now $|G|/G_1 = 12/3 = 4$ axes ϕ of rotations of order 3, and $|G|/|G_2| = 12/4 = 3$ axes of rotations of order 2. The nontrivial elements of G are exhausted by these $4(3 - 1) + 3(2 - 1) = 11$ rotations. Since the axes of the three rotations α, β, γ of order 2 are in the same class, these three elements are conjugate, and conjugation by some element δ of order 3 must permute them cyclically. Since $d_2 = 2$, some element of G must transpose the fixed points of α, and this can only be β or γ, say β. Then $\alpha\beta = \beta\alpha$, and $\gamma = \alpha\beta$, whence $V = \{1, \alpha, \beta, \gamma\}$ is a four-group, and $G = \langle \delta \rangle \cdot V$ semidirect product. (The remaining elements or order 3 in G are $\delta\alpha$, $\delta\beta$, $\delta\gamma$.)

We can recognize G as a rotation group of the sphere S as follows. The three rotations of order 2 have axes the three lines through the mid-points of opposite faces of an inscribed cube, and the four rotations of order 3 have axes the four diagonals of the cube.

We can realize G in M^+ by the following transformations:

$\alpha : z \mapsto -z$, fixed points $0, \infty$;

$\beta : z \mapsto 1/z$, fixed points $+1, -1$;

$\gamma : z \mapsto -1/z$, fixed points $+i, -i$;

$\delta : z \mapsto \dfrac{-z+i}{z+i}$, mapping $0, 1, i$ and $\infty, -1, -i$ cyclically, hence conjugating α, β, γ cyclically.

<u>Case 4</u>. Let $n = 3$. Since $d_i = 1$ implies $t_i \geqslant \frac{1}{2}$, while all $t_i \geqslant \frac{1}{4}$, we must have all $d_i = 2$. The equation $\Sigma\, t_i = 1 - \frac{1}{|G|}$ reduces to $\Sigma \frac{1}{p_i} = \frac{2}{|G|}$. Choosing notation with $p_1 \leqslant p_2 \leqslant p_3$, this has only the solutions $(p_1,p_2,p_3) = (2,3,3), (2,3,4), (2,3,5)$, with $|G| = 12, 24, 60$.

For G a group of rotations of S, these are realized as the groups of rotational symmetries of the tetrahedron, octahedron, and icosahedron.

The octahedral case, (2,3,4) can be realized in M^+ from the group $\langle \alpha, \beta, \gamma, \delta \rangle$ obtained above by adding a further generator $\varepsilon : z \mapsto iz$ to obtain the normalizer N(V) of V. We do not attempt to represent the tetrahedral and icosahedral groups in M^+, but remark only that they (as well as the others) can be obtained from rotation groups of S by stereographic projections.

THEOREM. <u>The finite subgroups of M^+ are isomorphic to the finite groups of rotations of the sphere</u> S.

7. Simplicity of M^+

M^+ is generated by products of two inversions. We first examine the structure of such a product.

THEOREM. Let $\alpha = \gamma_{c_1}\gamma_{c_2} \neq 1$, the product of inversion in two distinct inversive lines. If c_1 and c_2 are tangent, then α is parabolic. If c_1 and c_2 are not tangent, then α is loxodromic with real trace $t = a + d$. If c_1 and c_2 intersect, then $t^2 < 4$ and α is elliptic (conjugate to a rotation); if not, and $t^2 > 4$ then α is hyperbolic (conjugate to a dilation).

Proof. We use the observation that all points common to c_1 and c_2 are fixed points of α, and that any further fixed points must lie on the Euclidean line orthogonal to both (the line joining their centers if both are circles). It follows directly that if c_1 and c_2 are tangent, the only fixed point is their point of tangency.

Suppose now that c_1 and c_2 are not tangent. After conjugation we can suppose that c_1 is the unit circle and c_2 a line $x = k$ for some real $k \neq \pm 1$. From the fact that α maps 0 to ∞, ∞ to $2k$, and 1 to $2k - 1$ we find, by substitution, that the matrix for α has the form $M = \begin{pmatrix} 2k & -1 \\ 1 & 0 \end{pmatrix}$, with trace $t = 2k$. If M has eigenvalues λ and λ^{-1}, then $t = \lambda + \lambda^{-1}$, whence $\lambda = \dfrac{t \pm \sqrt{t^2 - 4}}{2}$. Now λ and $A = \lambda^2$ are real just in case $t^2 > 4$. If $t^2 < 4$, then λ is not real, but the condition that $\lambda + \lambda^{-1} = t$ be real implies that, for some θ, $\lambda = e^{i\theta}$ and $A = e^{2i\theta}$. Finally we note that c_1 and c_2 intersect just in case $|k| < 1$, that is, $t^2 = 4k^2 < 4$. □

COROLLARY. The nontrivial products of inversions in tangent inversive lines are precisely the parabolic elements of M^+. The products of inversions in inversive lines that are not tangent are precisely the loxodromic elements with real trace, that is, the elliptic and hyperbolic elements, of M^+.

Proof. This follows from the theorem using the fact established earlier that nontrivial elements with the same trace are conjugate.

Remark. Compare this with the Euclidean case, where every element of M^+ is a product of two reflections.

THEOREM. M^+ is generated by parabolic elements.

Proof. M^+ is generated by all products $\alpha = \gamma_{c_1}\gamma_{c_2}$ of two inversions. There always exists an inversive line c tangent to both c_1 and c_2. Now $\beta_1 = \gamma_{c_1}\gamma_c$ and $\beta_2 = \gamma_c\gamma_{c_2}$ are parabolic elements, and, since $\gamma_c^2 = 1$, $\alpha = \beta_1\beta_2$. □

THEOREM. M^+ is simple, that is, its only normal subgroups N are the trivial subgroup 1 and the whole group M^+.

Proof. Suppose $N \neq 1$. Since all parabolic elements are conjugate and N is normal, if N contains any parabolic element it contains all, and, since M^+ is generated by parabolic elements, $N = M^+$. Thus it suffices to show that N contains a parabolic element. Since $N \neq 1$, it contains some element $\alpha \neq 1$. If α is parabolic, then the argument above shows that $N = M^+$. Suppose then that α is loxodromic with fixed points P and Q. Let β be any parabolic element of M^+ with fixed point P. Then β^α is a parabolic element with fixed point P, and, since we have seen that $\alpha\beta \neq \beta\alpha$, $\beta^\alpha \neq \beta$. It follows that $\beta^{-1}\beta^\alpha$ is a parabolic element (with fixed point P). But $\beta^{-1}\beta^\alpha = \beta^{-1}\alpha^{-1}\beta^\alpha = (\alpha^\beta)^{-1}\alpha$, where α and α^β are in N, whence the parabolic element $\beta^{-1}\beta^\alpha$ is in N. □

Remarks. 1. This shows that there is no nontrivial decomposition of M^+ as a semidirect product, analogous to the decomposition $E^+ = E_0^+ \cdot T$.

2. The proof of this theorem contains the germ of the proof of a more general theorem, that PSL(n,F) is simple for all $n \geqslant 2$ and all fields F, except in the cases $n = 2$ and $F = Z_2$ or $F = Z_3$. The argument is closely analogous to the proof that the alternating group A_n is simple for $n \geqslant 5$.

Problems

Problem 1. Show that any function $f(z_1,z_2,z_3)$ of three points in $\underline{C}^*$ that is invariant under M assumes the same value for all triples of distinct points. Show that the cross ratio $[z_1,z_2,z_3,z_4] = \frac{z_1 - z_3}{z_1 - z_4} \cdot \frac{z_2 - z_4}{z_2 - z_3}$ of four distinct points (with the usual definitions regarding ∞) is invariant under M. Show that for three fixed distinct points a, b, c, and z, $z' \neq$ a, b, c, one has $[a,b,c,z] = [a,b,c,z']$ if and only if $z = z'$. Show that $[z_1,z_2,z_3,z_4]$ is real if and only if all four points line on a common inversive line. Show that $|\ell n(|[z_1,z_2,z_3,z_4]|)|$ is unchanged by any permutation of the four points.

Problem 2. Let H be a subgroup of a group G of permutations of a set Ω, and let β be any element of G. Show that if H has orbits $\Omega_1,\ldots,\Omega_t$ then H^β has orbits $\Omega_1\beta,\ldots,\Omega_t\beta$.

Problem 3. Show that $T_P \simeq \underline{C}^+$, the additive group of complex numbers, and that $M_P^+/T_P \simeq \underline{C}^x$, the multiplicative group of nonzero complex numbers. Show that, in fact, $M_P^+ = M_{P,Q}^+ \cdot T_P$, semidirect product of T_P by $M_{P,Q}^+$, for any $Q \neq P$, and describe the conjugation map from $M_{P,Q}^+$ into Aut T_P.

Problem 4. Let $F = [0,0'] \cup [A,A'] \cup [B,B']$, the union of three orthogonal diameters of a sphere S. Suppose that S is tangent to the complex plane $\underline{C}$ at the point 0. Show that, for suitable choice of the radius of S, the points 0, 0', A, A', B, B' project orthogonally (from 0') onto 0, ∞, +1, -1, +i, -i. Deduce that N(V) is isomorphic to the symmetry group of the figure F, hence to that of the octahedron with vertices 0, 0', A, A', B, B'.

Problem 5. In $G = PSL(2,F)$, for any field F, a transvection is a transformation conjugate to a translation $x \mapsto x + c$, $c \in F$. Show that G is generated by transvections and that, if $|F| > 3$, then G is simple. Show that $F = \underline{Z}_2, \underline{Z}_3$ are counterexamples.

Problem 6. Supply the details of the following discussion. We use the facts that M^+ is transitive on inversive lines and that $\gamma_c^\alpha = \gamma_{(c\alpha)}$. Let $\alpha = \gamma_{c_1}\gamma_{c_2}$, a product of two distinct inversions.

(1) If c_1 and c_2 are tangent we can transform them to two distinct parallel Euclidean lines in $\underline{C}$, whence α is parabolic and a Euclidean translation with real $t^2 = 4$.

(2) If c_1 and c_2 meet, we can transform them to a pair of intersecting Euclidean lines, whence α is elliptic and a Euclidean rotation with real $t^2 < 4$,

(3) Let c_1 and c_2 be disjoint. Then we can transform them to two Euclidean circles, one inside the other. If they are concentric, then α is hyperbolic, a Euclidean dilation with real $t^2 > 4$. If not, we can suppose that their centers lie on $\underline{R}$, whence the coefficients of γ_{c_1}, γ_{c_2}, and α are real, and t is real. We show that all fixed points of α must be real, whence $t^2 \geqslant 4$. If $t^2 > 4$, then α is hyperbolic, but it can happen that $t^2 = 2$ and α is parabolic.

(4) Since the general loxodromic transformation is conjugate to the product of a Euclidean rotation and a Euclidean dilation with the same center, every element of M^+ is a product of four inversions, and there exist elements that are not the product of fewer than four inversions.

References

Inversive geometry is discussed in many books on Euclidean geometry, for example, in Guggenheimer. For an easily readable discussion of inversion in connection with *Fuchsian groups*, the discontinuous subgroups of the group M^+ of all *Möbius transformations*, see Ford. For a very thorough and detailed account of these matters, see Beardon.

CHAPTER NINE: HYPERBOLIC GEOMETRY

1. The hyperbolic group and hyperbolic plane

We now examine the stabilizer M^+_ℓ in M^+ of an inversive line ℓ. Since M^+ is transitive on the set of all inversive lines, all stabilizers M^+_ℓ of inversive lines are conjugate within M^+, and, in this sense they are all geometrically equivalent under 'change of coordinates'. Two choices of the line ℓ present special advantages, the extended real line $\underline{R}^* = \underline{R} \cup \{\infty\}$ and the unit circle Γ. We shall use both choices, but begin by taking $\ell = \underline{R}^*$.

We define the <u>hyperbolic group</u> H to be $H = M^+_{\underline{R}^*}$.

THEOREM. $H \simeq LF(2,\underline{R}) = PSL(2,\underline{R})$.

<u>Proof</u>. We view $LF(2,\underline{R})$ as a subgroup of $LF(2,\underline{C}) = M^+$, comprising all transformations $\alpha : z \mapsto \dfrac{az + c}{bz + d}$, with $ad - bc = 1$, for which the coefficients a, b, c, d lie in $\underline{R}$. Clearly such a transformation α maps $\underline{R}^*$ to itself. For the converse, suppose that some α in M^+ maps $\underline{R}^*$ to itself. Just as $LF(2,\underline{C})$ is exactly triply transitive on $\underline{C}^*$, we see the $LF(2,\underline{R})$ is exactly triply transitive on $\underline{R}^*$. Thus $LF(2,\underline{R})$ contains some β such that $0\beta = 0\alpha$, $1\beta = 1\alpha$, $\infty\beta = \infty\alpha$. From the exact triple transitivity of $LF(2,\underline{C})$ we conclude that $\alpha = \beta$, in $LF(2,\underline{R})$. □

THEOREM. H <u>maps the upper half plane</u> $H = \{x + iy : y > 0\}$ <u>to itself</u>.

<u>Proof</u>. The line $\underline{R}^*$ divides $\underline{C}^*$ into the two disjoint connected components H and $\bar{H} = \{x + iy : y < 0\}$ of $\underline{C}^* - \underline{R}^*$. Therefore, each $\alpha \in H$,

mapping $\underline{R}^*$ to itself, must permute H and $\bar{H}$. We calculate

$$i\alpha = \frac{c + ai}{d + bi}\,\frac{d - bi}{d - bi} = \frac{(cd + ab) + (ad - bc)i}{d^2 + b^2} = \frac{(cd + ab) + i}{d^2 + b^2} = x + yi \text{ with}$$

$y > 0$. Thus α maps the point $i \in H$ to $i\alpha \in H$, and so must map H to itself. □

We define the hyperbolic plane, as an incidence plane, to have the set of points H, with hyperbolic lines the parts $\ell \cap H$ in H of all inversive lines ℓ orthogonal to $\underline{R}^*$, that is, of all Euclidean circles with center on $\underline{R}$ and all Euclidean lines perpendicular to $\underline{R}$. We shall later introduce a hyperbolic metric on H.

In practice one does not always distinguish sharply between ℓ and the part $\ell \cap H$ of ℓ in H. In particular, one often speaks of the ends (points at infinity) of the hyperbolic line $\ell \cap H$ as the two points in which ℓ meets $\underline{R}^*$. Henceforth we use the word 'line' to mean hyperbolic line, unless otherwise stated, and we use letters ℓ, ... for such lines.

THEOREM. *$\mathcal{H}$ maps lines to lines and is transitive on the set of all lines.*

Proof. Since $\mathcal{H} \subseteq \mathcal{M}^+$ and $\mathcal{M}^+$ preserves angles, an element α of $\mathcal{H}$, which maps $\underline{R}^*$ to itself, must map any inversive line orthogonal to $\underline{R}^*$ to some inversive line orthogonal to $\underline{R}^*$, that is, α maps hyperbolic lines to hyperbolic lines. To show transitivity, note that a hyperbolic line is uniquely determined by its two ends on $\underline{R}^*$. Since $\mathcal{H}$ is (more than) doubly transitive on $\underline{R}^*$, it contains α mapping the ends of a line ℓ to those of ℓ', and hence ℓ to ℓ'. □

THEOREM. *Every nontrivial element of $\mathcal{H}$ is either **elliptic** with exactly one fixed point in $\mathcal{H}$ **parabolic** with exactly one fixed point in $\underline{R}^*$, or **hyperbolic** with two fixed points in $\underline{R}^*$.*

Proof. This follows from the previous classification of elements α of $\mathcal{M}^+$ according to the trace $t = a + d$, noting now that, for $\alpha \in \mathcal{H}$, t must be real. We found that the fixed points of α were given by the formula $z = \frac{t \pm \sqrt{t^2 - 4}}{2}$. If α is parabolic, with $t^2 = 4$, then α has only one fixed point $z = \frac{t}{2}$ on $\underline{R}^*$. If α is loxodromic and $t^2 < 4$, then the fixed points are two conjugate nonreal complex numbers, of which only one lies in H. If α is loxodromic with $t^2 > 4$, then α has two distinct real fixed points. □

The discussion above illustrates one more abuse of language. In practice it is not necessary to distinguish between $\alpha \in \mathcal{H}$, technically a bijection of the points of $\underline{C}^*$, and its restriction α_H to H, an automorphism of the incidence plane H. In particular, we will speak of α as acting on the boundary $\partial H = \underline{R}^*$ of H.

Since $\mathcal{H}$ is more than doubly transitive on points of $\underline{R}^*$, we expect that it is more than simply transitive on hyperbolic lines, that is, that the stabilizer $\mathcal{H}_\ell$ in $\mathcal{H}$ of a hyperbolic line ℓ should be far from trivial.

THEOREM. *The stabilizer $\mathcal{H}_\ell$ in $\mathcal{H}$ of a line ℓ is exactly simply transitive on ℓ, and $\mathcal{H}_\ell$ is isomorphic to $\underline{R}^x_+$, the group of positive reals under multiplication.*

Proof. Since $\mathcal{H}$ is transitive on lines, it suffices to prove this assertion for any conveniently chosen line ℓ, and we choose to take $\ell = I = \{iy : y > 0\}$, the upper half of the imaginary axis. Now $\alpha \in \mathcal{H}_\ell$ must fix the two ends 0 and ∞ of ℓ, and thus be of the form $\alpha : z \mapsto a^2 z$ for real $a \neq 0$. It is clear that the group of all such α is isomorphic to $\underline{R}^x_+$, the multiplicative group of all positive reals, and acts exactly simply transitively on ℓ. □

COROLLARY. *$\mathcal{H}$ is transitive on the points of* H.

Proof. Let $P \neq P'$. From Euclidean geometry it is clear that there is a (unique) hyperbolic line ℓ containing P and P'. By the theorem, some $\alpha \in H_\ell$ carries P to P'. □

2. Solvable subgroups of H

The discussion of solvable subgroups of H follows exactly the same pattern as the earlier discussion of solvable subgroups of M^+. These groups turn out to be exactly the subgroups of groups H_P for $P \in H$, of elliptic elements together with 1, of groups H_P for $P \in \underline{R}^*$, and of groups $H_{\{P,Q\}}$ for distinct $P, Q \in \underline{R}^*$.

The discussion of finite subgroups of H also follows the same pattern as for M^+, and we find that all finite subgroups of H are subgroups of some H_P for $P \in H$, hence are cyclic.

To begin, let G be a nontrivial abelian subgroup of H. Suppose that G contains an elliptic element α, with fixed point $P \in H$, that is, α is in the abelian group H_P. If $\beta \in H$ and $\alpha^\beta \in H_P$, then β must fix P, that is, $\beta \in H_P$. In particular, if α and β commute, then $\beta \in H_P$, whence $G \subseteq H_P$. More generally, if $G^\beta = G$ then $\beta \in H_P$, that is, $N(G) \subseteq H_P$. It follows that $N(H_P) = H_P$.

Next suppose that G contains a parabolic element α with fixed point $P \in \underline{R}^*$. If $\alpha^\beta \in T_P$, then $\beta \in H_P$. In particular, if α and β commute, then $\beta \in T_P$, whence $G \subseteq T_P$. More generally, $N(G) \subseteq H_P$, whence $N(T_P) = H_P$. We note also that if K is a subgroup of H_P such that $K \cap T_P$, the subgroup of parabolic elements of H_P, is not trivial, then any β normalizing K must also normalize $K \cap T_P$, and hence lie in H_P. It follows that $N(H_P) = H_P$.

Finally suppose that G contains a hyperbolic element α with fixed points P and Q on $\underline{R}^*$. If $\alpha^\beta \in H_{P,Q}$ then $\beta \in H_{\{P,Q\}}$, that is, β either fixes both P and Q or interchanges them. Suppose that β interchanges P and Q. Without loss of generality we may suppose that $P = \infty$ and $Q = 0$.

Then we have $\alpha : z \mapsto Az$ for some real $A \neq 0, \pm 1$ and $\beta : z \mapsto \frac{B}{z}$ for real $B \neq 0$, and $\alpha\beta \neq \beta\alpha$. We conclude that $G \subseteq H_{P,Q}$, that $N(G) \subseteq H_{\{P,Q\}}$, whence $N(H_{P,Q}) = H_{\{P,Q\}}$. It is immediate that $N(H_{\{P,Q\}}) = H_{\{P,Q\}}$.

We recapitulate all these facts as a theorem.

THEOREM. (1) _The maximal abelian subgroups of_ H _are of three types: the group_ H_P _of (elliptic) elements fixing a point_ $P \in H$; _the group_ T_P _of all parabolic elements fixing a point_ $P \in \underline{R}^*$; _the group_ $H_{P,Q}$ _of (hyperbolic) elements fixing a pair_ P, Q _of distinct points of_ $\underline{R}^*$. _(Note: these groups also contain, of course_, 1.)

(2) If $1 \neq G \subseteq H_P$, $P \in H$, then $N(G) \subseteq H_P$;

if $1 \neq G \subseteq T_P$, then $N(G) \subseteq H_P$, and if $K \subseteq H_P$ with $K \cap T_P \neq 1$, then $N(K) \subseteq H_P$;

if $1 \neq G \subseteq H_{P,Q}$, then $N(G) \subseteq H_{\{P,Q\}}$.

(3) For $P \in H$, $N(H_P) = H_P$;

for $P \in \underline{R}^*$, $N(T_P) = H_P$ and $N(H_P) = H_P$;

for $P, Q \in \underline{R}^*$, $N(H_{P,Q}) = H_{\{P,Q\}}$ and $N(H_{\{P,Q\}}) = H_{\{P,Q\}}$.

Reversing the indexing above, we now suppose we have a chain of groups $G_1 \vartriangle G_2 \vartriangle \cdots \vartriangle G_n = G$, contained in H, where G_1 is a nontrivial abelian group and each G_i is normal in G_{i+1} for $1 \leqslant i < n$. If $G_1 \subseteq H_P$, $P \in H$, then, by the theorem, $N(G_1) \subseteq H_P$, whence $G_2 \subseteq H_P$, and, repeating this argument, $G \subseteq H_P$. If $G_1 \subseteq T_P$, then $G_2 \subseteq N(G_1) \subseteq H_P$, and, since $1 \neq G_1 \subseteq G_2 \cap T_P$, $G_3 \subseteq N(G_2) \subseteq H_P$; repeating the argument gives $G \subseteq H_P$. Finally, if $G_1 \subseteq H_{P,Q}$ then $G_2 \subseteq H_{\{P,Q\}}$ and, repeating, $G \subseteq H_{\{P,Q\}}$.

THEOREM. _The only solvable subgroups of_ H _are subgroups of groups of type_ H_P, $P \in H$; H_P, $P \in \underline{R}^*$; _and_ $H_{\{P,Q\}}$, $P, Q \in \underline{R}^*$. For $P \in H$, $H_P \simeq \mathrm{Aut}^+\Gamma$ is abelian. For $P \in \underline{R}^*$, H_P has an abelian normal subgroup $T_P \simeq \underline{R}^+$ with quotient group $H_P/T_P \simeq \underline{R}_+^x$ abelian. For distinct $P, Q \in \underline{R}^*$, $H_{\{P,Q\}}$ has an abelian normal subgroup $H_{P,Q} \simeq \underline{R}_+^x$ with quotient group $H_{\{P,Q\}}/H_{P,Q} \simeq C_2$

abelian.

THEOREM. All finite subgroups of $\mathcal{H}$ are subgroups of $\mathcal{H}_P$ for some $P \in H$, and hence are finite cyclic groups.

Proof. We use a method similar to that used in studying the finite subgroups of $\mathcal{M}^+$. Let G be a finite subgroup of $\mathcal{H}$. Then each nontrivial element α of G generates a finite cyclic group $\langle \alpha \rangle$ which, since $\langle \alpha \rangle$ is abelian, must be contained in some group $\mathcal{H}_P$ for $P \in H$. In short, each nontrivial element α of G is elliptic with a single fixed point P in H. If also $\beta \in G$, then $\alpha^\beta \in G$ has fixed point $P\beta$. Thus G permutes the set Ω of fixed points of nontrivial elements of G. Since $G - 1 = \bigsqcup_{P \in \Omega} (G_P - 1)$, disjoint union, we have $|G| - 1 = \sum_{P \in \Omega} (|G_P| - 1)$. Since all the $G_{P'}$ for P' in the orbit PG of P under G are conjugate and hence have the same order, we can collect equal terms in this sum to obtain $|G| - 1 = \Sigma\, |PG| \cdot (|G_P| - 1)$, summed over all orbits PG in Ω under G. We now use the relation $|G| = |PG| \cdot |G_P|$ to rewrite this as $|G| - 1 = \sum \frac{|G|}{|G_P|}(|G_P| - 1)$, and now divide through by $|G|$ to obtain $1 - \frac{1}{|G|} = \Sigma\, t_P$ where $t_P = \frac{|G_P| - 1}{|G_P|}$ depends only on the orbit PG. But now each $t_P \geqslant \frac{1}{2}$ while $\Sigma\, t_P < 1$, summed over all orbits PG. This implies that there is only one orbit PG, and $1 - \frac{1}{|G|} = t_P = 1 - \frac{1}{|G_P|}$, whence $|G| = |G_P|$, and $G = G_P \subseteq \mathcal{H}_P$.

3. Incidence and angle

In studying the action of the group $\mathcal{H}$ on H it is sometimes advantageous to consider $\mathcal{H}$ as a subgroup of a larger group $\tilde{\mathcal{H}}$ containing reflections in all hyperbolic lines of H.

We define $\tilde{\mathcal{H}} = \mathcal{M}_{\underline{R}^*}$, the stabilizer of $\underline{R}^*$ in $\mathcal{M}$.

THEOREM. The group $\tilde{\mathcal{H}}$ is generated by all inversions in hyperbolic lines of H, and $\mathcal{H} = \tilde{\mathcal{H}}^+$, the subgroup of $\tilde{\mathcal{H}}$ generated by all products of two such inversions. In particular, if $\alpha : z \mapsto -\bar{z}$ is inversion in the (upper half of) the imaginary axis, then $\tilde{H} = \langle \alpha \rangle \cdot H$, the semidirect product of H by the group $\langle \alpha \rangle$ of order 2.

Proof. The proof is routine and is left as an exercise. □

We now state three theorems about H as an incidence plane.

THEOREM. If P and Q are distinct points of H, then there is a unique hyperbolic line ℓ containing both P and Q.

Proof. Let e be the Euclidean line through P and Q. If e is perpendicular to $\underline{R}$, then $\ell = e \cap H$ is a hyperbolic line through P and Q, and is clearly the only such line. If e is not perpendicular to $\underline{R}$, then the Euclidean perpendicular bisector of the segment [P,Q] meets $\underline{R}$ in a point C. If c is the circle with center C passing through P and Q, then c is orthogonal to $\underline{R}$, whence $\ell = c \cap H$ is a hyperbolic line through P and Q, and clearly the only such line. □

THEOREM. If ℓ and ℓ' are hyperbolic lines, then they have at most one point in common.

Proof. Let $\ell = c \cap H$ and $\ell' = c' \cap H$ where c and c' are inversive lines orthogonal to $\underline{R}^*$. Then c and c' have at most two points in common, and, if they have two points of $\underline{C}^*$ in common, at most one lies in H. □

THEOREM. If ℓ is a line and P a point not on ℓ, then there are infinitely many lines ℓ' through P that do not meet ℓ.

Proof. We may suppose that $\ell = I = \{yi : y > 0\}$ and that $P = a + bi$ with $a, b > 0$. For every x such that $0 \leqslant x \leqslant a$ there is a line ℓ' through P with x as one end that does not meet ℓ. □

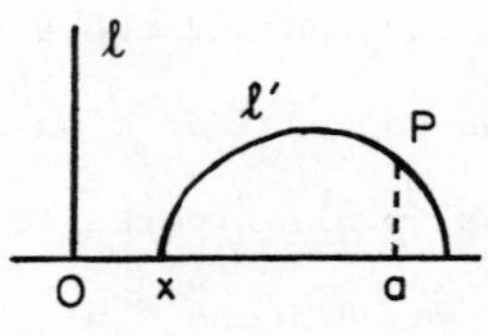

Remark. This theorem states that H is noneuclidean in the sense that it fails to satisfy Euclid's Parallel Postulate that, through each point

P not on a line ℓ, there is exactly one line ℓ' parallel to ℓ. Because of the abundance of lines ℓ' not meeting a given line ℓ, we hesitate to call them parallel. Sometimes the term parallel is used for two lines with a common end on $\underline{R}^*$, that is, with a common point at infinity; in the example used in the proof, there are two such lines through P, the limiting cases for $x = 0$ and ℓ' tangent to ℓ at 0, and for $x = a$, where ℓ' is a Euclidean parallel to ℓ with a common end at the point ∞ of $\underline{R}^*$.

THEOREM. *If* P *is a point*, ℓ *any line, and* θ *is any angle*, $0 < \theta < \pi$, *then there exists a unique line* ℓ' *through* P *meeting* ℓ *with the angle from* ℓ' *to* ℓ *equal to* θ.

Proof. We use the same figure as above, but now take $x < 0$ and permit $a = 0$. As x recedes along $\underline{R}$ from 0, the angle θ varies continuously in the range $0 < \theta < \pi$. □

COROLLARY. *If* ℓ *is any line and* P *any point, then there is a unique line* ℓ' *through* P *that is perpendicular to* ℓ.

THEOREM. *If* ABC *is any triangle in* H, *then the sum of the interior angles is less than* π.

Proof. We emphasize that we are speaking of a hyperbolic triangle, whose sides are segments of the hyperbolic lines AB, BC, CA passing through pairs of vertices.

To prove this theorem we use for the first time our other 'model' for the hyperbolic plane. Let $\Gamma = \{z : |z| = 1\}$ be the unit circle. Since M^+ is transitive on inversive lines, M^+ contains a transformation δ carrying $\underline{R}^*$ to Γ. In fact, we can take $\delta : z \mapsto \frac{z - i}{z + i}$ which maps 0 to -1, 1 to i, and ∞ to 1. Since also δ maps i to 0, it maps the upper half plane H to the interior D of Γ. Further, it maps all hyperbolic lines in H to the intersection with D of Euclidean lines and circles orthogonal to Γ. Thus we can replace H by D and H by the group H^δ, which in fact consists of all $\alpha : z \mapsto \frac{az + \bar{b}}{bz + \bar{a}}$ for a, b $\in \underline{C}$ such that $a\bar{a} + b\bar{b} = 1$. In practice one does not always distinguish between these two models, thinking of H with its group H and D with its group H^{δ} as two different coordinatizations of the same geometric object.

To return to our proof, in the model D the three lines AB, BC, CA are circles orthogonal to Γ (or one of them can be a diameter), and comparison with the Euclidean triangle with the same vertices A, B, C gives the desired inequality. □

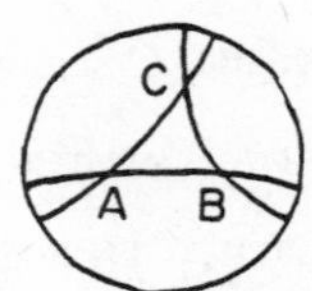

THEOREM. Let $\alpha, \beta, \gamma > 0$ be angles such that $\alpha + \beta + \gamma < \pi$. Then there exists a triangle with α, β, γ as interior angles (in cyclic order in the positive sense), and any two such triangles are congruent, that is, one is the image of the other under some element of H.

Proof. We continue to use the model D. It will suffice to show that if A = 0, there is a triangle ABC with angles α, β, γ, and that this triangle is unique apart from rotation about A = 0.

We begin with a Euclidean circle c with some center D. Let $\delta = \pi - (\alpha + \beta + \gamma)$, whence $0 < \delta < \pi$. Let DB and DC be two radii of c meeting at D with an angle δ. Let ℓ_B and ℓ_C be the lines through B and C making angles of β and γ with the shorter arc of c between B and C, Since $\delta + \beta + \gamma = \pi - \alpha < \pi$, ℓ_B and ℓ_C must meet at a point A to form a quadrilateral DCAB with interior angles $\delta, \gamma, \alpha, \beta$. Thus the curvilinear triangle ABC has interior angles α, β, γ. After a uniquely chosen trnaslation, we may suppose that A = 0.

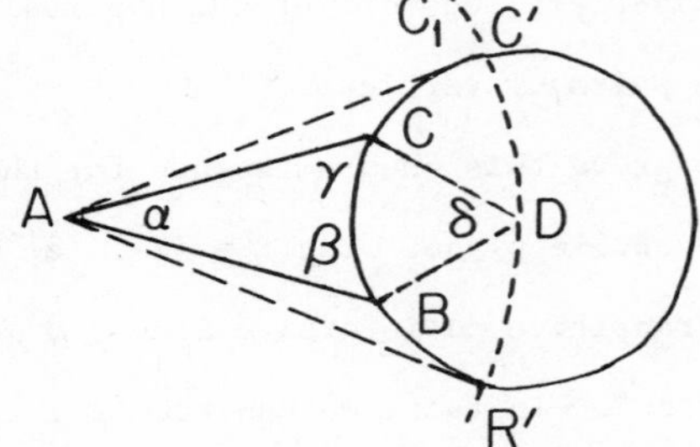

Let c_1 be the unique circle with center $A = 0$ that is orthogonal to c, meeting it at points B', C'. Since AB' and AC' are tangent to c, while AB and AC meet c in angles β, γ where $0 < \beta,\gamma < \pi$, the points B, C are nearer to A than the corresponding points B', C', hence are inside the circle c_1. After a uniquely determined dilation with center 0, we may suppose that $c_1 = \Gamma$. Now ABC is a hyperbolic triangle in D with angles α, β, γ. It is clear from the construction that ABC is unique up to rotation about $A = 0$. □

<u>Remark</u>. It is natural to call two hyperbolic triangles directly similar if they have the same interior angles, taken in cyclic order in the positive sense, and to call two triangles congruent if one is the image of the other under an element of *H*. We have shown that two directly similar triangles are always congruent. This is in sharp contrast with Euclidean geometry. We have not yet discussed hyperbolic length, but it is clear that any reasonable definition of hyperbolic length would require congruent segments to have the same length. Under such a definition we could conclude that if corresponding angles of two triangles are equal, then corresponding sides have the same length.

<u>Remark</u>. We did not emphasize that the vertices A, B, C of the triangle ABC all lie in the hyperbolic plane H. In fact, if one or more of them lies on $\partial H = \underline{R}^*$, with interior angle now 0, the theorem remains valid. The proof as given works if one or two vertices are on $\underline{R}^*$, while the case that all three are on $\underline{R}^*$ is trivial.

4. Combinatorial definition of area

We have put off discussing distance and related metric concepts in the hyperbolic plane as long as possible, proving as much as possible by purely combinatorial arguments. It is fairly clear that to discuss the length of a more or less arbitrary curve, or the area of a general region, differential concepts are needed; these will be given later. However, we are now in a position to give a very simple definition of the area of a region bounded by a simple closed polygon, and we turn to such a definition.

Let R be a region in H bounded by a simple closed polygon ∂R, whose sides are segments of hyperbolic lines. There is no need to exclude the case that some of the vertices lie on the boundary $\partial H = \underline{R}^*$ of H. Let α_V be the interior angle at a vertex V; if $V \in H$, then $0 < \alpha_V < 2\pi$, while if $V \in \underline{R}^*$, then $\alpha_V = 0$. As we traverse ∂R in the positive sense, in passing through the vertex V our direction, measured in the positive sense, changes by $\pi - \alpha_V$. This prompts us to make the following definition.

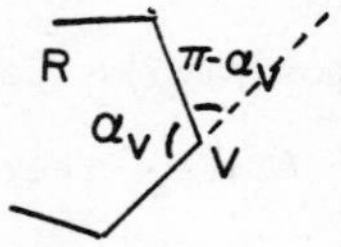

The <u>curvature</u> $\kappa(\partial R)$ of the boundary ∂R of R is $\Sigma(\pi - \alpha_V)$, summed over all vertices of R.

If R were a region bounded by a Euclidean polygon, the corresponding definition would always yield $\kappa(\partial R) = 2\pi$. However, we have seen that if R is bounded by a hyperbolic triangle, then $\kappa(\partial R) = 3\pi - (\alpha + \beta + \gamma) > 2\pi$. If the triangle is very small, say contained in a small neighborhood of $\underline{0}$ in the model D, its sides are nearly straight, whence $\alpha + \beta + \gamma$ is nearly π, and $\kappa(\partial R)$ is only slightly larger than 2π. If the triangle is very large, in the extreme case if all three vertices lie on $\underline{R}^*$, then α, β, γ are nearly 0 and $\kappa(\partial R)$ is nearly 3π. This is typical: we shall see that, for a hyperbolic polygonal region, always $\kappa(\partial R) > 2\pi$, with $\kappa(\partial R)$ near to 2π for small regions, and unboundedly large for larger regions.

We define the <u>area</u> A(R) of the polygonal region R to be

$A(R) = \kappa(\partial R) - 2\pi$.

There are two basic properties that we expect area, thus defined, to have. First, it should be invariant under the group H: if $\alpha \in H$ then $A(R\alpha) = A(R)$; this follows immediately from the definition. Second, we expect it to be 'additive', in the sense of the following theorem.

THEOREM. <u>If a polygonal region</u> R <u>is the nonoverlapping union of two polygonal regions</u> R_1 <u>and</u> R_2, <u>then</u> $A(R) = A(R_1) + A(R_2)$.

<u>Proof</u>. The hypothesis implies that the intersection of R_1 and R_2 is a simple polygonal path p on the boundaries of both. By inserting new vertices V with $\alpha_V = \pi$, which does not alter the definitions of $A(R_1)$, $A(R_2)$, $A(R)$, we may suppose that p runs from a common vertex V_1 to a common vertex V_2. Consider the sum $\Sigma(\pi - \alpha_V^{(1)}) + \Sigma(\pi - \alpha_V^{(2)})$, the first sum over all vertices of R_1 and the second over all vertices of R_2. Clearly the terms arising from vertices interior to p will cancel out. At V_1 the interior angle in R is $\alpha_{V_1} = \alpha_{V_1}^{(1)} + \alpha_{V_1}^{(2)}$, the sum of the interior angles in R_1 and R_2, whence $\pi - \alpha_{V_1} = (\pi - \alpha_{V_1}^{(1)}) + (\pi - \alpha_{V_1}^{(2)}) - \pi$. Similarly at V_2, whence $\Sigma(\pi - \alpha_V^{(1)}) + \Sigma(\pi - \alpha_V^{(2)}) = \Sigma(\pi - \alpha_V) + 2\pi$. It follows that $A(R_1) + A(R_2) = A(R)$. □

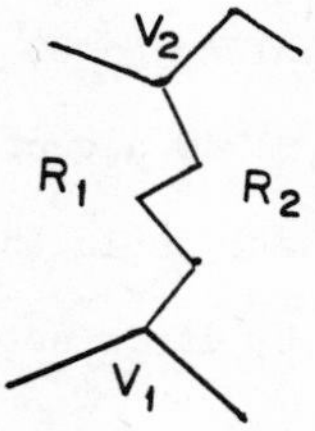

5. The hyperbolic metric

We now examine the question of an invariant distance function on H, and, with it, the derived definitions of length of a curve and area of a region. These ideas are important, but one not in the line of our present development, whence we only sketch them somewhat informally.

A metric on set U is a function $d : U \times U \to \underline{R}$ with the following properties:

(1) for all P, Q in U, $d(P,Q) \geqslant 0$, with $d(P,Q) = 0$ just in case $P = Q$;

(2) $d(P,Q) = d(Q,P)$;

(3) $d(P,R) \leqslant d(P,Q) + d(Q,R)$ (the triangle inequality).

A metric on U defines a topology on U, with a basis of open sets all balls $B_r(P) = \{Q : d(P,Q) < r\}$, for all $P \in U$ and $r > 0$. If U is already given as a topological space, one wants this new topology to agree with the given topology on U, which amounts to imposing the further condition

(4) $d(P,Q)$ is a continuous function of P and Q.

If, further, as in the case at hand, of $U = H$, the concept of line is defined in U, it is natural to require that these lines be geodesics in the sense that

(5) if Q lies between P and R on a line ℓ, then
$d(P,R) = d(P,Q) + d(Q,R)$.

Finally, in our case, that $U = H$, we require that the metric be invariant under H, that is,

(6) $d(P\alpha,Q\alpha) = d(P,Q)$ for all $P, Q \in H$ and $\alpha \in H$.

Clearly, if a metric d satisfies all these conditions, and a second metric d' is defined by setting $d'(P,Q) = kd(P,Q)$ for some $k > 0$, then d' also satisfies these conditions. We shall prove the following.

THEOREM. There exists an invariant metric d on H (satisfying conditions (1)-(6)), and d is unique up to multiplication by some $k > 0$.

Proof. If P, Q lie on some line ℓ, then some α in H maps ℓ to $I = \{iy : y > 0\}$. Now (6) requires that $d(P,Q) = d(P\alpha,Q\alpha)$, whence d is entirely determined by its values for $P, Q \in I$.

Let $P = ip$, $Q = iq \in I$, thus with $p, q > 0$. Now H_P consists of all transformations $z \mapsto Az$, $A > 0$, whence (6) requires that $d(ip,iq) =$

d(iAp,iAq). Thus d(P,Q) depends only on the ratio q/p. Now (2) implies that d(P,Q) depends only on $|\ell n(q/p)|$, say $d(ip,iq) = \phi(|\ell n(p/q)|)$. If $R = ir$, then (5) requires that $\phi(|\ell n(p/r)|) = \phi(|\ell n(p/q)|) + \phi(|\ell n(q/r)|)$, that is, that for all $x, y > 0$, $\phi(x + y) = \phi(x) + \phi(y)$. Finally, condition (4), that d be continuous, implies that ϕ is continuous, whence it follows that, for some $k > 0$, $\phi(x) = kx$ for all $x > 0$. Thus, if d satisfies the given conditions, then, for some $k > 0$, $d(ip,iq) = k\cdot|\ell n(q/p)|$ for all $q, p > 0$.

We normalize by taking $k = 1$, and $d(ip,iq) = |\ell n(p/q)|$. It is, with one exception, trivial to check that this d satisfies all the given conditions. The exception is condition (3), which involves three points not necessarily on the same line; we defer the proof that d satisfies this condition.

We now seek a <u>hyperbolic differential of arc length</u>, that is a differential $d_H z$, depending only on z and dz, such that if K is any curve (for which the integral exists), then the length of K is given by the formula

$$\ell(K) = \int_K d_H z.$$

We require that L(K) be invariant under H, that is that $\ell(K\alpha) = \ell(K)$ for all α in H, and that it agree with our previous definition of distance on a line: that $\ell([P,Q]) = d(P,Q)$.

To find an invariant differential $d_H z$, we first calculate that, for $\alpha : z \mapsto \frac{az + c}{bz + d}$, a, b, c, c $\in \underline{R}$, $ad - bc = 1$, the 'Euclidean' (ordinary) differential dz transforms according to the formula $d(z\alpha) = \frac{dz}{(bz + d)^2}$. A second calculation shows that the imaginary part Im z = y, for $z = x + iy$, transforms according to the formula $\mathrm{Im}(z\alpha) = \frac{\mathrm{Im}\ z}{|bz + d|^2}$. We conclude that $d_H z = \frac{|dz|}{\mathrm{Im}\ z}$ is invariant: $d_H(z\alpha) = d_H z$, whence $\ell(K\alpha) = \ell(K)$.

To check this against our formula for distance we let $P = ip$, $Q = iq$, $0 < p < q$, whence $[P,Q]$ consists of all $z = iy$ for $p \leqslant y \leqslant q$. Now our formula for L gives

$$\ell(P,Q) = \int_p^q \frac{dy}{y} = |\ell n(p/q)| = d(P,Q).$$

We now prove the triangle inequality, in the following form.

THEOREM. _Let_ P, Q, R _be points in_ H. _Then_

$$d(P,R) \leqslant d(P,Q) + d(Q,R)$$

with equality only if R _is contained in the segment_ [P,R].

Proof. We shall in fact prove a stronger result. Let C be a piecewise differentiable curve from P to R in H, with hyperbolic length $\ell(C)$; then $d(P,R) \leqslant \ell(C)$, with strict inequality provided the slope of C is bounded away from that of PR on some arc of C.

We take $P = ip$ and $R = ir$, with $0 < p < r$. We may decompose the interval $J = [P,R]$ into subintervals J_k such that, as iy runs through J_k the points $z = \phi(y) + iy$ run through an arc C_k of C, where ϕ is differentiable for $iy \in J_k$. Then, on each C_k, $|dz| = \sqrt{\phi'(y)^2 + 1}\, dy \geqslant dy$, whence $\ell(C_k) \geqslant \ell(J_k)$. Summing gives $\ell(C) \geqslant \ell(J) = d(P,R)$. If C contains points $z = x + iy$ for $y \notin J$, we clearly have strict inequality. If the slope of C is bounded on some arc, which we may suppose to be some C_k, then $|\phi'(y)|$ is bounded away from 0 on C_k, say $|\phi'(y)| > b > 0$, whence, on C_k, $|dz| \geqslant \sqrt{b^2 + 1}\, dy$ and $\ell(C_k) \geqslant \sqrt{b^2 + 1}\ \ell(J_k) > \ell(J_k)$, giving $\ell(C) > \ell(J)$. □

From the hyperbolic differential of arc, $\frac{dz}{y}$, we obtain $\frac{dxdy}{y^2}$ as hyperbolic differential of area, and hence, for the area of a region R, the formula $A(R) = \iint_R \frac{dxdy}{y^2}$.

To show that this agrees with our combinatorial definition, it is enough to verify it for triangles, where again there is no need to exclude triangles with one or more vertices on the boundary ∂H. The figure shows that every triangle is the difference of one disjoint union of two triangles each with a vertex at ∞, that is, with two vertical sides and another such triangle.

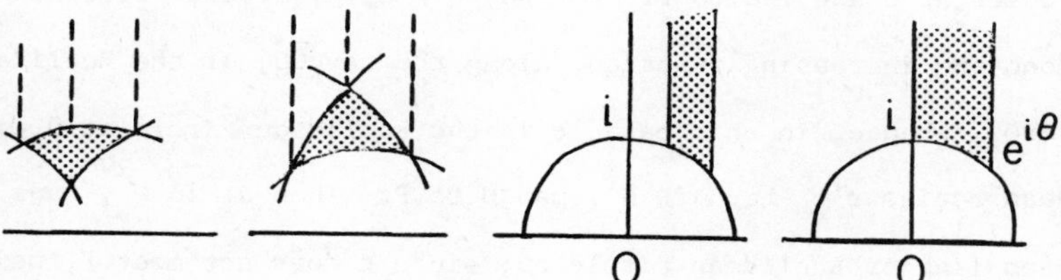

Thus it suffices to consider triangles with two vertical sides and the remaining side on a circle, which we may suppose to be Γ. Since each such triangle is the difference of two such triangles with one side on I, we are reduced to treating the following case: the triangle has vertices ∞, i, and $e^{i\theta}$ for some θ, $0 < \theta < \pi/2$. The formula for A(R) now gives

$$A(R) = \iint_R \frac{dxdy}{y^2} = \int_{x=0}^{x=\cos\theta} \int_{y=\sqrt{1-x^2}}^{y=\infty} \frac{dxdy}{y^2}$$

$$= \int_{x=0}^{x=\cos\theta} \frac{dx}{\sqrt{1-x^2}} = -\arccos\theta \Big|_{x=0}^{x=\cos\theta}$$

$$= \frac{\pi}{2} - \theta = \pi - \left(\theta + \frac{\pi}{2} + 0\right),$$

in accordance with the combinatorial formula.

6. Two more theorems

Hyperbolic circles do not play a role in our discussion, but deserve a note nonetheless. It is natural to define the <u>hyperbolic circle</u> with center P and radius $r > 0$ to consist of all points Q such that the hyperbolic distance $d(P,Q) = r$.

THEOREM. The hyperbolic circle c with center P and hyperbolic radius $r > 0$ is a Euclidean circle, but in general with some other Euclidean center P' and radius $r' > 0$.

Proof. We first consider the hyperbolic circle in the disc model D with center at 0 and radius r. For any $Q \neq 0$, hyperbolic distance $d(0,Q)$ is a monotone increasing function, along the ray $\overrightarrow{0Q}$, of the Euclidean distance $|PQ|$, whence, in this case, c is the Euclidean circle at 0 with some Euclidean radius r'. Let $\alpha \in H^{\delta}$, map 0 to P. Then α, in M^{+}, maps c to a Euclidean line or Euclidean circle $c\alpha$; since c does not meet Γ, neither does $c\alpha$, and $c\alpha$ is a Euclidean circle. Note that 0α will not ordinarily be the Euclidean center of $c\alpha$. □

The next theorem will play an important role in Chapter 10.

THEOREM. Let P, Q be distinct points of H. Then the locus of points X such that $d(P,X) = d(Q,X)$ is a hyperbolic line, the perpendicular bisector of the segment $[P,Q]$.

Proof. It is easy to see that $[P,Q]$ contains a unique midpoint M such that $d(P,M) = d(Q,M)$. By an earlier theorem, there is a unique line ℓ through M perpendicular to PQ. Let $\rho \in \tilde{H}$ be inversion in ℓ. If $X \in \ell$, then, since $X\rho = X$ and $P\rho = Q$, $d(P,X) = d(QX)$.

For the converse, suppose that a point X is equidistant from P and Q; we may suppose that $X \notin PQ$. Let ℓ' be the unique line through X bisecting the angle between PX and QX, and let ρ' be inversion in ℓ'. Since ρ' fixes X and maps PX to QX, and $d(P,X) = d(Q,X)$, ρ' maps P to Q. If ℓ' meets PQ in a point M', then ρ' fixes M' and maps the triangle XPM' to XQM'. It follows that M' is the midpoint of $[P,Q]$ and that ℓ' is perpendicular to PQ. Thus $\ell' = \ell$ and $X \in \ell$. □

Problems

Problem 1. Prove the theorem that $\tilde{H}$ is the split extension of H by $<\alpha>$ for α reflection in the imaginary axis.

Problem 2. The solvable subgroups of H are of more interest as exceptions to various theorems about Fuchsian groups than in their own right. These elementary groups G are sometimes characterized as those leaving invariant some finite nonempty set X of points in $\bar{H} = H \cup \underline{R}^*$. Explore this hypothesis for small values of $|X| = 1, 2, 3, \ldots$.

Problem 3. We have seen that the range of areas of hyperbolic triangles is exactly all A in the interval $0 < A \leqslant \pi$. What about the areas of polygons with n sides?

Problem 4. Find a formula for the area of a regular p-gon with interior angles $\frac{2\pi}{q}$ for positive integers p, q such that $\frac{1}{p} + \frac{1}{q} < \frac{1}{2}$. For fixed p, what happens as we increase q?

Problem 5. Let c be the hyperbolic circle with center P and radius $r > 0$, consisting of all z such that $d(P,z) = r$. What are its center P' and radius r' as Euclidean circle?

Problem 6. Let there be given two distinct points A and B in H and two positive reals a and b. How many points C are there such that $d(A,C) = b$ and $d(B,C) = a$? Prove that if two triangles have corresponding sides equal, then they are congruent under $\tilde{H}$. What if the triangles have two pairs of corresponding sides and the included angles equal?

Problem 7. For z a point on the upper half of the unit circle, express d(i,z) in terms of the slope of the line at -1 through z.

Reference

Among the various references cited elsewhere, the book of Beardon is probably the most relevant to this chapter.

CHAPTER TEN: FUCHSIAN GROUPS

1. Fundamental regions

A Fuchsian group is a discontinuous subgroup of the hyperbolic group $\mathcal{H}$. We recall that a subgroup G of $\mathcal{H}$ is discontinuous if and only if, given a point P in H and a disc D in H, there are only finitely many images $P\alpha$ of P, for α in G, in the disc D. The theory of Fuchsian groups is very highly developed and currently very active. However, in this concluding chapter we concentrate on a few important examples, developing the basic theory just enough to enable us to understand these examples.

Our study of Fuchsian groups is based on the concept of a fundamental set. This is a closed subset Δ of H such that H is the nonoverlapping union of the images $\Delta\alpha$, $\alpha \in G$. That is, H is the union of the $\Delta\alpha$, and, for $\alpha_1 \neq \alpha_2$, any point common to $\Delta\alpha_1$ and $\Delta\alpha_2$ lies on the boundary of both. To prove the existence of a manageable fundamental set, we need the following theorem.

THEOREM. *If G is a Fuchsian group and F is the set of fixed points of elliptic elements of G, then F is a discrete subset of H.*

Proof. Suppose that F is not discrete, that is, for some point P there exists a sequence of elliptic elements α_n of G with fixed points $P_n \neq P$ such that $\lim d(P_n,P) = 0$. Then, since $d(P\alpha_n,P) \leqslant d(P\alpha_n,P_n) + d(P_n,P) = d(P\alpha_n,P_n\alpha_n) + d(P_n,P) = 2d(P_n,P)$, it follows that $\lim d(P\alpha_n,P) = 0$. But this implies that every disc D containing P also contains infinitely many $P\alpha_n \neq P$, which contradicts the fact that G is discontinuous. □

We remark that this theorem is the easier half of a theorem of Fenchel and Nielsen; the other half says that if F is discrete and contains more than two elements, then G is discontinuous.

We use this theorem only to conclude that H contains points P not fixed by any nontrivial element of G. Given such a point P, we define the Dirichlet region $\Delta = \Delta(P)$ as follows:

$$\Delta = \{Q : d(Q,P) \leqslant d(Q,P\alpha) \text{ for all } \alpha \text{ in } G\}.$$

THEOREM. *If* G *is a Fuchsian group and* P *is a point of* H *that is not fixed by any nontrivial element of* G, *then the Dirichlet region* $\Delta = \Delta(P)$ *is a fundamental set for* G.

Proof. (1) Since $\Delta\alpha$ is the set of all Q that are at least as near to $P\alpha$ as to any other $P\beta$, and, by discontinuity, there are only finitely many $P\beta$ within any given distance of Q, it is clear that every $Q \in H$ is in some $\Delta\alpha$.

(2) If $Q \in \Delta\alpha_1 \cap \Delta\alpha_2$, then $d(Q,P\alpha_1) = d(Q,P\alpha_2)$ and Q lies on the locus of all R such that $d(R,P\alpha_1) = d(R,P\alpha_2)$, which, by a theorem of the preceding chapter, is a (hyperbolic) line ℓ. Now the segment $[P\alpha_1,Q]$ is contained in $\Delta\alpha_1$ while any point of the line from $P\alpha_1$ through Q that is beyond Q will be nearer to $P\alpha_2$ than to $P\alpha_1$, and hence not in $\Delta\alpha_1$. It follows that Q is on the boundary of $\Delta\alpha_1$. □

We note that Δ determines a Dirichlet tessellation T of H with faces the $\Delta\alpha$, $\alpha \in G$, which are congruent, but not ordinarily regular polygons.

In general, a Fuchsian group G will have many fundamental regions, not all of which will be Dirichlet regions. For example, let Δ be a polygonal fundamental region, and Δ_0 a region contained in Δ and meeting

$\partial\Delta$ in a segment of some side of Δ; then, for a certain $\alpha \in G$, the closure of $(\Delta - \Delta_0) \cup \Delta_0\alpha$, obtained from Δ by removing Δ_0 and replacing it by $\Delta_0\alpha$, will be another fundamental region Δ'.

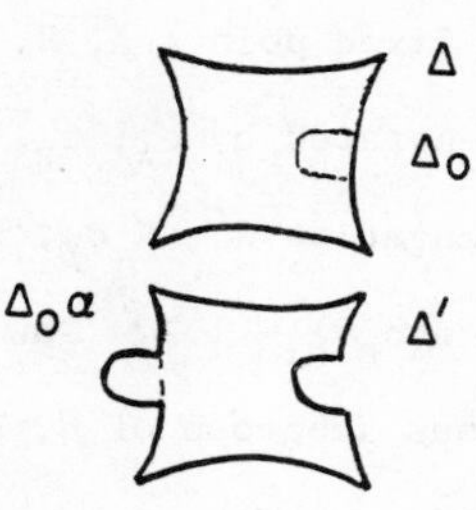

We describe, without proofs (which will be supplied later), some fundamental regions for Fuchsian groups, which are in fact Dirichlet regions for suitable choice of the point P.

<u>Example 1</u>. The <u>modular group</u> G, which arises in algebra, analysis, and number theory, can be described as the group G generated by an elliptic element α of order 2 with fixed point i and an elliptic element β of order 3 with fixed point $\omega = \frac{1 + i\sqrt{3}}{2}$. (Their product $\gamma = \alpha\beta$ is the map $\gamma : z \mapsto z + 1$.) If $P = iy$, $y > 1$, the resulting Dirichlet region is as shown in the figure, a triangle of infinite extent, with vertices ω, ω^2, and ∞.

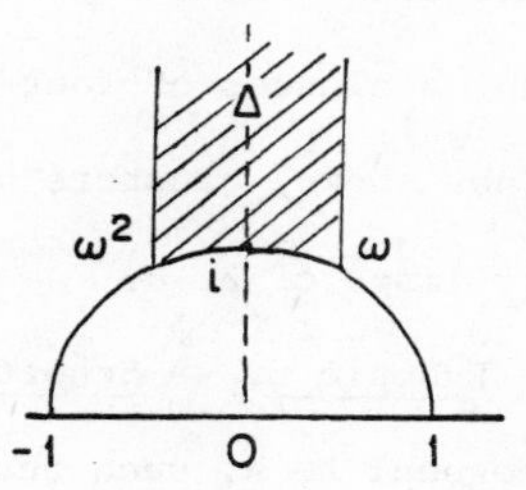

<u>Esample 2</u>. A <u>triangle group</u>. Let $\tilde{\Delta}$ be the region bounded by a triangle ABC, with interior angles $\frac{2\pi}{a}$, $\frac{2\pi}{b}$, $\frac{2\pi}{c}$ at the vertices A, B, C. We know that such a triangle exists if a, b, c are positive integers such that $\frac{1}{a} + \frac{1}{b} + \frac{1}{c} < \frac{1}{2}$. Let α, β, γ be inversions in the three sides, opposite A, B, C, and $\tilde{G}$ the subgroup of $\tilde{H}$ generated by α, β, γ. Just as in the Euclidean case we know that $\tilde{G}$ has fundamental region $\tilde{\Delta}$, hence is a discontinuous subgroup of $\tilde{H}$, and that $\tilde{G}$ has a presentation

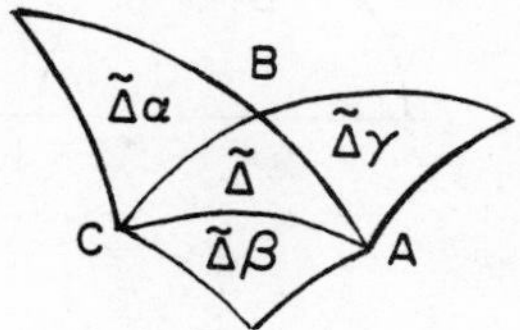

$$\tilde{G} = \langle \alpha, \beta, \gamma : \alpha^2 = \beta^2 = \gamma^2 = (\beta\gamma)^a = (\gamma\alpha)^b = (\alpha\beta)^c = 1 \rangle.$$

Let $\sigma_A = \beta\gamma$, $\sigma_B = \gamma\alpha$, $\sigma_C = \alpha\beta$, elliptic elements of orders a, b, c with fixed points A, B, C. Then $G = \tilde{G}^+$ is generated by σ_A, σ_B, σ_C, and has a presentation $G = < \sigma_A, \sigma_B, \sigma_C : \sigma_A^a = \sigma_B^b = \sigma_C^c = \sigma_A\sigma_B\sigma_C = 1 >$. Now G is a discontinuous subgroup of H, hence a Fuchsian group, and we know on general principle that the union of $\tilde{\Delta}$ and an abutting replica of $\tilde{\Delta}$, say $\Delta = \tilde{\Delta} \cup \tilde{\Delta}\alpha$, is a fundamental region for G. This is illustrated in the figure, where we have taken c = 4. Then there is, as shown, a cluster of four copies of Δ about the vertex C, and there will be (not shown) clusters of a faces about every image of A and of b about every image of B.

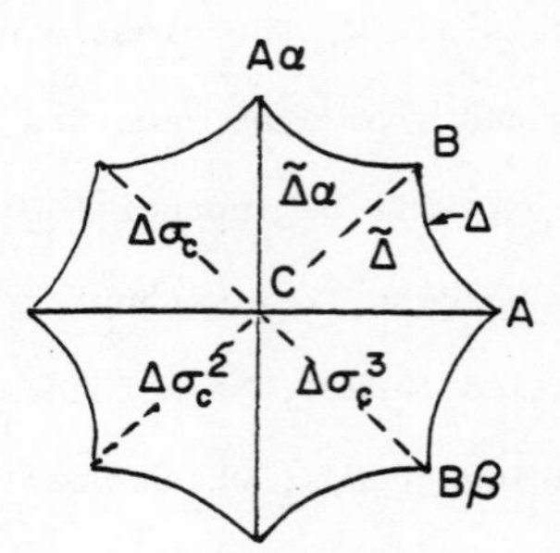

Example 3. A Schottky group. Let ℓ_1, ℓ_2, ℓ_3 be three circles orthogonal to $\underline{R}$, such that the open half discs D_1, D_2, D_3 enclosed by them (and segments of $\underline{R}$) are disjoint. Let ρ_1, ρ_2, ρ_3 be the three inversions, in $\tilde{H}$, in the hyperbolic lines ℓ_1, ℓ_2, ℓ_3.

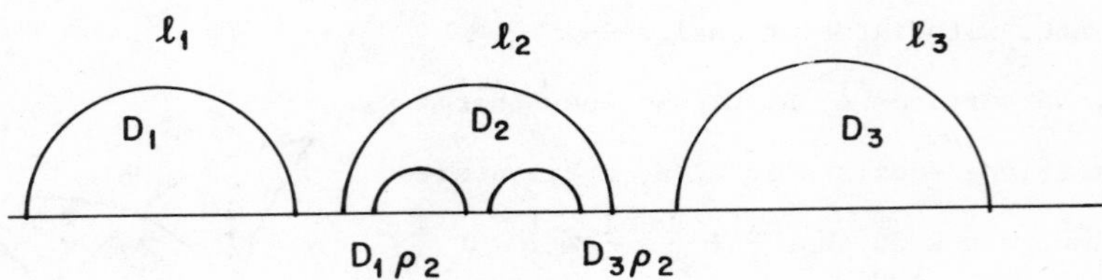

Let $\tilde{G}$ be the subgroup of $\tilde{H}$ generated by ρ_1, ρ_2, ρ_3. One can show that $\tilde{G}$ has a fundamental region $\tilde{\Delta} = H - (D_1 \cup D_2 \cup D_3)$, and a presentation $G = < \rho_1, \rho_2, \rho_3 : \rho_1^2 = \rho_2^2 = \rho_3^2 = 1 >$.

Let $G = \tilde{G}^+$, generated by $\alpha_1 = \rho_1\rho_2$, $\alpha_2 = \rho_2\rho_3$, $\alpha_3 = \rho_3\rho_1$: since $\alpha_1\alpha_2\alpha_3 = 1$, G is in fact generated by α_1 and α_3. One can conclude that the Fuchsian group G has a fundamental region $\Delta = \tilde{\Delta} \cup \tilde{\Delta}\rho_2 =$ $H - (D_1 \cup D_3 \cup D_1\rho_2 \cup D_3\rho_2)$, and a presentation $G = \langle \alpha_1, \alpha_3 : \emptyset \rangle$ with no defining relations, that is, that G is a free group with basis $\{\alpha_1, \alpha_3\}$.

2. Geometry of the fundamental region

We assume that Δ is a Dirichlet region for a Fuchsian group G, and show that Δ is in a suitable sense a polygonal region, that is, that its boundary is either a closed polygon or a union of polygonal arcs (possibly with infinitely many sides) joining points of $\underline{R}$.

A subset U of H is *convex* if, whenever two points P and Q are in H, then the segment [P,Q] joining them is in H.

THEOREM. *Δ is closed and convex.*

Proof. By definition Δ is the intersection of the closed half spaces $H_\alpha = \{Q : d(Q,P) \leqslant d(Q,P\alpha)\}$, for all nontrivial α in G. We have seen that the boundary ∂H_α of H_α, the locus of Q such that $d(Q,P) = d(Q,P\alpha)$, is a hyperbolic line ℓ_α. It follows that each H_α is closed and convex, whence their intersection Δ is also closed and convex. □

For our next theorem we shall need the following.

LEMMA. *If D is any disc in H, then only finitely many $\Delta\alpha$ have non-empty intersection with D.*

Proof. Since every disc D is contained in some larger disc with hyperbolic center at P, it suffices to prove this for D a disc with center at P. Let D have radius $r > 0$. If $Q \in D \cap \Delta\alpha$, then $d(Q,P\alpha) \leqslant d(Q,P) < r$, whence $d(P,P\alpha) \leqslant d(P,Q) + d(Q,P\alpha) < 2r$. By discontinuity, this is possible for only finitely many $\Delta\alpha$. □

DEFINITION. A finite interval of a line ℓ is a segment [P,Q] for some P, Q on ℓ; it is degenerate if P = Q, whence [P,Q] = {P}. An infinite interval of ℓ is either a doubly infinite interval, ℓ itself, or a simply infinite interval, one of the two components of $\ell - \{P\}$ for some point P on ℓ. The endpoints of an interval s are the endpoints of its closure $\bar{s}$ in $\bar{H} = H \cup \underline{R}^*$; an endpoint is finite if it is in H and infinite if it is in $\underline{R}^*$.

THEOREM. The boundary $\partial\Delta$ of Δ is the union of (possibly infinitely many) nondegenerate intervals s. These intervals meet at most in common endpoints, and such an endpoint belongs to at most two intervals, with each finite endpoint belonging to exactly two intervals.

Proof. Let $Q \in \partial\Delta$. Then any disc D with center Q contains points R not in Δ, hence in some $\Delta\alpha$, $\alpha \neq 1$. Since $Q \in \Delta$ and $R \in \Delta\alpha$, the locus ℓ_α of points equidistant from P and Pα, which we know to be a line, must intersect [Q,R], hence must meet D. Since only finitely many such ℓ_α can meet D, by taking D smaller we can suppose that the finite set of ℓ_α meeting D all pass through Q. Thus Q is in $s_\alpha = \ell_\alpha \cap \Delta$, which as a convex subset of ℓ_α must be an interval, possibly degenerate. We have shown that $\partial\Delta$ is the union of intervals $s_\alpha = \ell_\alpha \cap \Delta$, possibly degenerate.

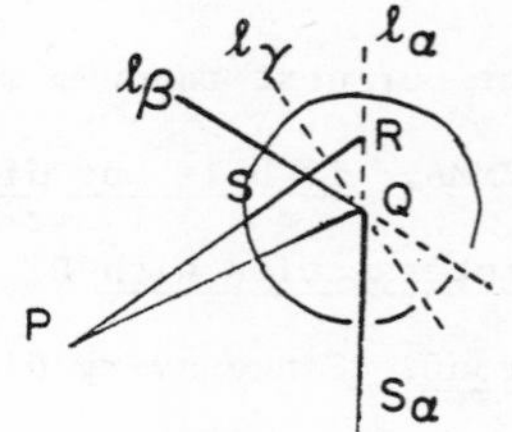

Now suppose that Q is an endpoint of s_α. Then D contains points R on ℓ_α 'beyond Q', that is, not on s_α. Let ℓ_β be that one among the ℓ_β through Q other than ℓ_α that is nearest to P. For R near enough to Q, [P,R] meets ℓ_β at a point S inside D. Now [P,S] meets no other $\ell_\delta \neq \ell_\beta$, since then ℓ_δ would have to meet either [P,Q] or [Q,S] in a point other than Q, which, in either case, contradicts the hypotheses. It follows that $S \in \partial\Delta$, whence $[S,Q] \subseteq \Delta \cap \ell_\beta = s_\beta$.

Without assuming that s_α was nondegenerate, we have shown that the endpoint Q of s_β is also an endpoint of the nondegenerate interval s_β. The same argument with s_β now in the role of s_α now shows that the endpoint Q of s_β is also the endpoint of some other nondegenerate interval s_α. From the convexity of Δ it now follows that any third ℓ_γ through Q can meet Δ only in the degenerate interval $s_\gamma = \{Q\}$. □

The sides of Δ are the nondegenerate intervals making up $\partial\Delta$. A side is of the form $s = \Delta \cap \Delta\alpha$, $\alpha \neq 1$. The finite vertices are the endpoints P in H of sides. If a side s is an infinite interval, its closure in $\bar{H} = H \cup \underline{R}^*$ will have one or two infinite vertices, that is, endpoints on $\underline{R}^*$. Although Δ is a closed subset of H, it may not be closed in $\bar{H}$, and we write $\bar{\Delta}$ for its closure in $\bar{H}$; now $\bar{\Delta}$ may have sides at infinity, that is, intervals of $\underline{R}^*$, that are not sides of Δ.

In keeping with our geometric point of view, we will now tacitly confine attention to the case that $\bar{\Delta}$ has only finitely many sides. With this assumption, suppose also that $\partial\bar{\Delta}$ is a simple closed polygon. This will always be the case if Δ is compact, that is, if Δ is contained in some disc. We now permit Δ to have vertices on $\underline{R}^*$ but no sides on $\underline{R}^*$, that is, we require that all sides be segments of hyperbolic lines. If V is any infinite vertex, as we traverse $\partial\bar{\Delta}$ in the positive sense there will be one side ending at V and another s' beginning at V; since s and s' are segments of hyperbolic lines, they are both perpendicular to $\underline{R}^*$ at V, and hence meet at an interior angle of 0. We conclude that Δ has finite area. Finally, if $\bar{\Delta}$ has sides at infinity, it is easy to see that Δ has infinite area.

Before leaving our discussion of the geometry of Δ, we want to make a minor but convenient modification of our definitions. Suppose that α

is an elliptic element of G of order $n \geqslant 3$. Let F be the fixed point of α and D any disc with center F; then, for any point Q of D, $Q \neq V$, the n images $Q\alpha^k$ must be distinct points of D. It follows that F must be a vertex of some $\Delta\beta$. If an elliptic element α of G of order $n = 2$ has fixed point F, the same reasoning shows that F must lie on a side of some $\Delta\beta$, but F need not be a vertex. We modify our definitions to count such points F as vertices.

Precisely, suppose that s is a side of Δ, as constructed above, between Δ from $\Delta\alpha$ where $\alpha^2 = 1$. Then α, interchanging Δ and $\Delta\alpha$, must map s to itself, while fixing the midpoint F of S. We now modify Δ by counting each such point F as a vertex; the side s is now replaced by two sides, the two segments into which F divides s.

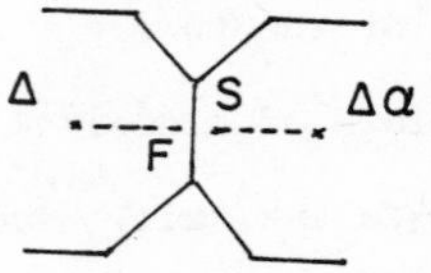

With this definition, each side s has the form $s = \Delta \cap \Delta\alpha$ unless $\alpha^2 = 1$, while, if $\alpha^2 = 1$, $\Delta \cap \Delta\alpha$ is the union $s \cup s\alpha$ of two sides s and $s\alpha$.

We suppose henceforth that Δ has been modified in this way.

3. Digression: elements of finite order

Any nontrivial element of finite order in M^+ must be elliptic. Conversely, an elliptic element of a Fuchsian group G must be of finite order: if elliptic $\alpha \in G$ fixes V, and $Q \neq V$ lies in a disc D at V, the $Q\alpha^k$ are distinct for distinct α^k, whence α must have finite order.

Suppose that Δ has only finitely many finite vertices V. It follows that the group G_V of elements fixing V is finite, whence there are, in all, only finitely many elements fixing some vertex of Δ. Now let α be any nontrivial element of G of finite order. Then α is elliptic with fixed point a vertex of some $\Delta\beta$, and its conjugate $\alpha^{\beta^{-1}}$ fixes a vertex of Δ. This proves the following.

THEOREM. <u>If</u> G <u>a Fuchsian group possessing a fundamental region</u> Δ <u>with only finitely many finite vertices, in particular, if</u> Δ <u>is compact, then there are only finitely many conjugacy classes of elements of finite order in</u> G.

One can show similarly, with a little more trouble, that if V is an infinite vertex of Δ where two sides s and $s\alpha$ meet, then α is parabolic with fixed point V, and that all parabolic elements of G are conjugate to elements fixing such an infinite vertex V of Δ. If Δ has only finitely many sides, one concludes that G contains only finitely many conjugacy classes of maximal parabolic subgroups.

4. The Cayley tessellation

If Δ is a Dirichlet region for a Fuchsian group G, then the $\Delta\alpha$, $\alpha \in G$, are the faces of a tessellation T of H, the <u>Dirichlet tessellation</u> of H. The faces $\Delta\alpha$ are all congruent, but not ordinarily regular polygonal regions. In fact, if all the vertices of Δ lie in H, then the faces are bounded by simple closed polygons; otherwise the closure of a face in $\bar{H} = H \cup \underline{R}^*$ is a polygonal region in $\bar{H}$, possibly with infinitely many sides.

We shall define a <u>Cayley tessellation</u> T^*, dual to T, whose 1-skeleton (set of edges and vertices) is a Cayley graph in the customary sense. For vertices of T^* we take all the images $P\alpha$, $\alpha \in G$, of the base point P of the Dirichlet region Δ. If s is a side, that is an edge in T, and $s \subseteq \Delta\alpha \cap \Delta\beta$, $\alpha \neq \beta$, we choose an edge s^* in T^* joining $P\alpha$ and $P\beta$ across s; these s^* can be chosen disjoint except at their endpoints. This defines the Cayley graph in H, and the faces of T^* are the components of the complement. If V is a vertex of T, then the edges s^* of T^* for all sides s of T at V form the boundary ∂V^* of a region V^* containing V; then V^* is a face of T^* and all faces of T^* are of this form. (If $V \in H$, then ∂V^* is a finite simple closed polygon. If $V \in \underline{R}^*$, then ∂V^* is an infinite polygonal path, which, together with a segment of $\underline{R}^*$, bounds the closure of V^* in $\bar{H}$.)

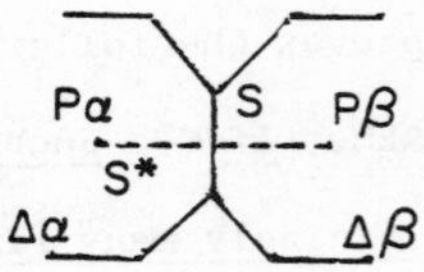

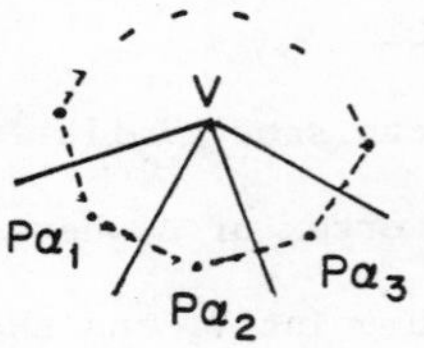

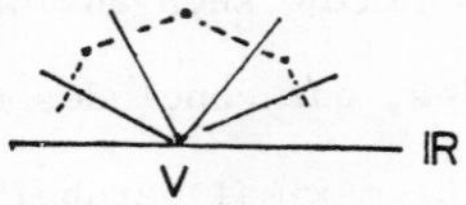

Let S be the set of sides of Δ. If $t \in S$, then $t \subseteq \Delta \cap \Delta\alpha$ for some unique $\alpha \neq 1$ in G, and we define a map $\phi : S \to G$ by setting $t\phi = \alpha$. Now $t\alpha^{-1} \subseteq \Delta\alpha^{-1} \cap \Delta$, whence $\bar{t} = t\alpha^{-1} \in S$ and $\bar{t}\phi = \alpha^{-1}$. We define a map $\eta : S \to S$ by setting $t\eta = \bar{t} = t(t\phi)^{-1}$. Evidently η is an involution without fixed points: $\bar{\bar{t}} = t$ and $t \neq \bar{t}$, even if $\alpha = \alpha^{-1}$.

If F is the free group with basis S, then ϕ extends uniquely to a homomorphism $\phi : F \to G$. We shall show presently that ϕ maps F onto G, and from this obtain a presentation of the form $G = \langle S:R \rangle$.

We use the map $\phi : S \to G$ to define a map from the paths in T^* into G, which we also denote by ϕ. If s^* is any edge of T^*, then $s = t\alpha$ for unique $t \in S$ and $\alpha \in G$, and we define $s^*\phi = t\phi$. If $p = s_1^* \ldots s_n^*$ is any path in T^*, we define $p\phi = (s_n^*\phi)\ldots(s_1^*\phi)$; note the reversed order of subscripts.

For $0 \leqslant k \leqslant n$, write $p_k = s_1^* \ldots s_k^*$ and $\alpha_k = p_k\phi$, where p begins at α_0. Assume inductively for $k < n$ that $\alpha_k = \alpha_0(p_k\phi)$ and that p_k runs from

α_0 to α_k. Now $s_{k+1} \subseteq \Delta\alpha_k \cap \Delta\alpha_{k+1}$ is the image $t_{k+1}\alpha_k$ of some side $t_{k+1} \subseteq \Delta \cap \Delta\beta_k$ where $\beta_k\alpha_k = \alpha_{k+1}$. Since $s_{k+1}{}^*\phi = t_{k+1}\phi = \beta_k$, $\alpha_{k+1} = (s_{k+1}{}^*\phi)\alpha_k = (s_{k+1}{}^*\phi)(p_k\phi) = p_{k+1}\phi$, we have completed the inductive proof of the following.

LEMMA. <u>If</u> p <u>is a path in</u> T^* <u>from</u> $P\alpha_0$ <u>to</u> $P\alpha_n$, <u>then</u> $\alpha_n = \alpha_0(p\phi)$. <u>In particular</u>, p <u>is a closed path if and only if</u> $p\phi = 1$.

Let $t_1,\ldots,t_n \in S$. It is customary to interpret the <u>word</u> $w = t_1\ldots t_n$ to mean either the sequence of the t_i or their product in F. (This ambiguity is similar to the use of the notation $a_1 + a_2 + \ldots$ to mean either the infinite series with terms a_i or the sum of the series, if one exists.) Given the sequence w of $t_1,\ldots,t_n$ and any vertex $P\alpha_0$ of T^*, it is clear that there is a unique path $p = s_n^*\ldots s_1^*$ at α_0 in T^* such that all $s_i^*\phi = t_i$, whence $p\phi = (t_1\ldots t_n)\phi = w\phi$. We denote this path p by p(w). We have shown in particular the following.

LEMMA. <u>For any</u> w <u>in</u> F, w <u>is in the kernel</u> N <u>of the map</u> $\phi : F \to G$, <u>that is</u>, $w\phi = 1$, <u>if and only if</u> w(p) <u>is a closed path (for all</u> $P\alpha_0$).

5. The Poincaré presentation

We now establish a presentation $G = \langle S{:}R \rangle$ that can be derived directly from combinatorial properties of the tessellation T.

PROPOSITION. ϕ <u>maps</u> F <u>onto</u> G.

<u>Proof</u>. We must show that $S\phi$ generates G. Let $\alpha \in G$. If p is a path in T^* from P to $P\alpha$, then, by the above, $p\phi = \alpha$, and α is a product of factors from $S\phi$. It suffices to know that such a path p exists. This is equivalent to knowing that T^* is connected, in the sense that every pair of vertices in T^* is connected by some path in T^*. We take this as obvious. (A proof, in terms of T rather than T^*, was indicated earlier.) □

It remains to find a subset R of F such that the normal closure of F is N, the kernel of ϕ. Then we will have a presentation $G = \langle S:R \rangle$.

The elements of R will be of two types. For the first type, let $t \in S$. Then $p = t^*(\bar{t}^*(t\phi))$ is a closed path, indeed of a rather trivial nature: it is a U-turn, consisting of an edge t^* followed first in one direction and then back again in the opposite direction. In fact, $w(p) = t\bar{t}$ and $(w(p))\phi = p\phi = 1$. We define R_1 to be the set of all words $t\bar{t}$ for $t \in S$. (The set $S\phi$ is highly redundant in that it contains together with each generator $x = t\phi$ also $x^{-1} = \bar{t}\phi$; the relators in R_1 compensate by giving the relations $(t\phi)^{-1} = \bar{t}\phi$.)

The second type of relator is associated with the vertices of Δ that lie in H. Let V be such a vertex and V^* the corresponding face of T^* containing V. If p_V is the closed path at P going once around ∂V^* in the positive sense, then $(w(p_V))\phi = p_V\phi = 1$. We define R_2 to be the set of all words $w(p_V)$ for vertices V of Δ.

THEOREM. G _has a presentation_ $G = \langle S : R_1 \cup R_2 \rangle$.

Proof. We have chosen $R = R_1 \cup R_2 \subseteq N$, and it remains to show that N is the normal closure of R, generated by all conjugates of (powers of) elements from R. For this, let w be a word (representing an element) in N, and let p be a closed path in T^* such that $w(p) = w$. We will show how to modify p in such a way that corresponding w is modified by multiplication with a conjugate of an element of R, and how, by a succession of such modifications, p can be reduced to trivial path at some point, and thus w to the trivial element 1 of F.

Suppose first that p contains a U-turn, that is, for some paths p' and p", $p = p'up"$ where $u = s_i^* s_{i+1}^*$ and $s_i = t\alpha_i$, $s_{i+1} = \bar{t}\alpha_{i+1}$ for some $t \in S$. Then $w = w(p")(t\bar{t})w(p') = w(p'p")(t\bar{t})^{w(p')}$. Since $(t\bar{t})^{w(p')}$ is a conjugate of an element of R_1, we can replace p by p'p". Thus we are free, in modifying p, to delete (or insert) U-turns at will.

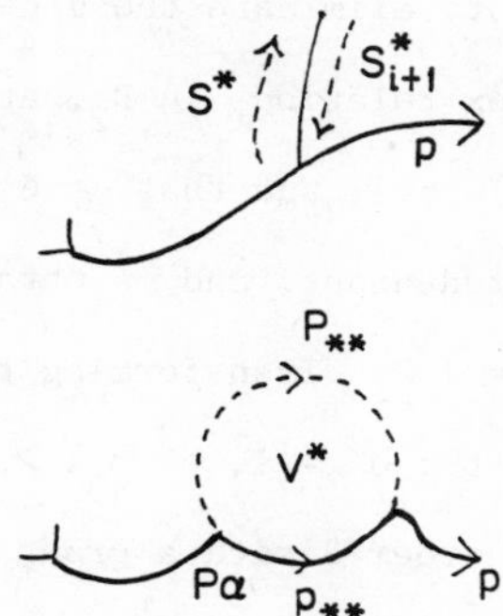

We next suppose that a segment p_* of $p = p'p_*p''$ is an arc beginning at a point $P\alpha$ of ∂V^*, of length $n \geqslant 0$ (if $n = 0$, p_* reduces to the point $P\alpha$). Now the path around ∂V^* at $P\alpha$, in one sense or the other, has the form $p_V = p_*p_{**}^{-1}$. Now $V = V_1\gamma$ for some vertex V_1 of Δ and some $\gamma \in G$, whence $w(p_V) = (w(p_{V_1})^u)^{\pm 1}$ is a conjugate of an element of R_2 where $u \in F$ with $u\phi = \gamma$. We find that $w(p)\ w(p_V)^{w(p')} = w(p'p_{**}p'')$. Thus we are permitted to modify p by replacing the arc p_* going around the boundary of V^* on one side of V^* by the complementary arc p_{**} going around the other side.

We have noted earlier that it is rather obvious intuitively that by modifications of the sorts above the closed path p can be reduced to a point. By rather obvious induction, removing simple closed loops, the proof reduces to the case that p is a simple closed path. By a second induction, on the number of faces V^* enclosed by p, we can modify p successively by going around the other side of some V^* enclosed by p, and thus reduce p to a point. □

<u>Example</u>. Let G be the modular group with fundamental region Δ as before. Then Δ, as shown, has vertices A, B, C, and ∞. (Note that B is the fixed point of an elliptic element of order 2.) There are four sides t_1, t_1, t_2, t_2 as shown, and $S = \{t_1, \bar{t}_1, t_2, \bar{t}_2\}$. As

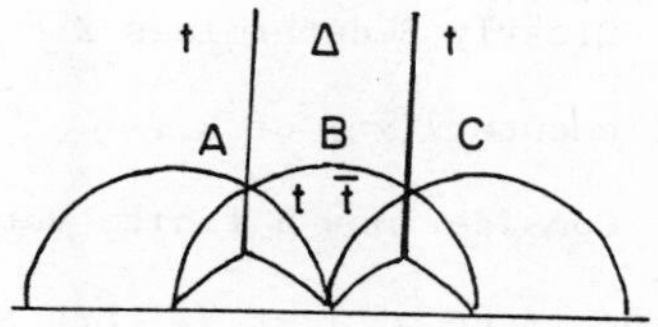

noted earlier, we can use the vertex relations from R_1, that $\bar{t}_1 = t_1^{-1}$ and $\bar{t}_2 = t_2^{-1}$ to eliminate the generators $\bar{t}_1$, $\bar{t}_2$ and replace S by $S' = \{t_1, t_2\}$. The vertex relation, in R_2, at B gives $t_2^2 = 1$. The vertex relation at A is $(t_1t_2)^3 = 1$, and that at C is $(t_1t_2)^3 = 1$. Evidently the third relation is redundant, and we obtain a presentation $G = \langle t_1, t_2 : t_2^2 = 1, (t_1t_2)^3 = 1 \rangle$. Transforming to generators $a = t_2$ and $b = t_1t_2$ gives $G = \langle a, b : a^2 = 1, b^3 = 1 \rangle$, whence $G \simeq C_2 \star C_3$, the free product of a group of order 2 with a group of order 3.

6. Combinatorial description of the presentation

It is implicit in the arguments above that the Poincaré presentation is obtainable from very limited knowledge about Δ. We now make this explicit.

We have already used the involution $\eta : S \to S$, where, if $t \subseteq \Delta \cap \Delta\alpha$, then $t\eta = t\alpha^{-1}$.

We define a second function θ by setting $t' = t\theta$ in case t' begins at the vertex V where t ends, in traversing $\partial\Delta$ in the positive sense. If $\partial\Delta$ is a polygon, then θ is, like η, a permutation of S; more generally, it may be only a function defined on a subset of S.

We define a third function ω by setting $t\omega$ equal to the interior angle of Δ at V in case t ends at V.

It will be shown that these three functions suffice to determine the presentation $G = \langle X:R \rangle$.

Clearly S determines $X = S$, and the involution η determines the set of elements $x(x\eta)$ of R_1.

Consider now a finite vertex V, and let $\Delta\alpha_1, \ldots, \Delta\alpha_n$, in cyclic order, be the faces of T at V. Let $s_i = t_i\alpha_i$ be the edge between $\Delta\alpha_i$ and $\Delta\alpha_{i+1}$. Now $t_i\eta = t_i\alpha_{t_i}^{-1}$, and we have seen that $\alpha_{i+1} = \alpha_{t_i}\alpha_i$, whence $(t_i\eta)\alpha_{i+1} =$

$t_i\alpha_i = s_i$. Since s_i follows s_{i+1} on $\partial\Delta\alpha_{i+1}$, and $s_i = (t_i\eta)\alpha_{i+1}$, $s_{i+1} = t_{i+1}\alpha_{i+1}$, it must be that $t_i\eta$ follows t_{i+1} on $\partial\Delta$, that is, that $t_i = t_{i+1}\theta\eta$. (We remark that, since V is a finite vertex, θ is defined on t_{i+1}.)

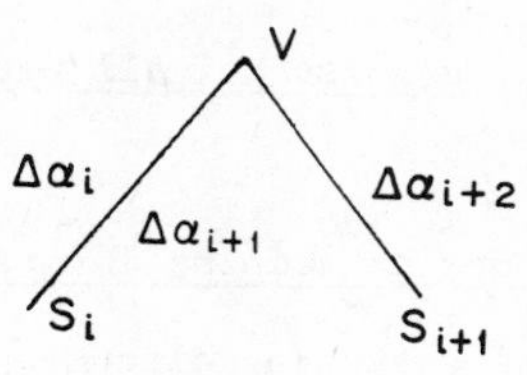

Define $\sigma = \theta\eta$.

Now $w(p_{V,\alpha_n}) = x_n \ldots x_1 = t_n \ldots t_1$, where each $t_i = t_{i+1}\sigma$, and $t_n = t_1\sigma$. Writing $x_n = x$, $w = w(p_{V,\alpha_n}) = x(x\sigma)(x\sigma^2)\ldots(x\sigma^{n-1})$, while $x\sigma^n = x$. If p is the least positive integer such that $x\sigma^p = x$, it may be that $p < n$; in this case p must divide $n = pq$, $q \geqslant 1$. Now $x, x\sigma, x\sigma^2, \ldots, x\sigma^{p-1}$ are all distinct, and $x\sigma^p = x$; in the usual language of permutations (although σ may not be a permutation of all of S), the cyclically ordered set $(x, x\sigma, \ldots, x\sigma^{p-1})$ is a <u>cycle</u> of σ.

With each cycle $c = (y_1, \ldots, y_p)$ of σ we associate the word $w(c) = y_1 \ldots y_p$. In this notation, we have $w = w(c)^q$. It is routine to show that this reasoning is reversible: if elements $x_i = t_i$ form a cycle c of σ of length p, then for some V, γ and some q, $w(P_{V,\gamma}) = w(c)^q$.

The sum of the angles at V is 2π, whence $\sum_1^n t_i\omega = 2\pi$. If p and q are as above, this gives $q(t_1\omega + \ldots + t_p\omega) = 2\pi$. For the cycle $c = (t_1, \ldots, t_p)$, we define $c\omega = t_1\omega + \ldots + t_p\omega$. Then we have $c\omega = \frac{2\pi}{q}$ for some integer $q \geqslant 1$. This determines $q = q(c)$, and hence the relator $w(P_{V,\gamma}) = w(c)q(c)$.

We summarize this result.

THEOREM. *For* $X = S$, *with* η, θ, ω *and* $\sigma = \theta\eta$, *let* C *be the set of cycles of* σ. *For each cycle* c *in* C, *let* w(c) *and* $c\omega$ *be defined as above. Then each* $c\omega = \frac{2\pi}{q}$ *for some integer* $q = q(c) \geqslant 1$. *Now* G *has a presentation*

$$G = \langle X : R_1 \cup R_2 \rangle$$

<u>where</u> R_1 <u>consists of all words</u> $x(x\eta)$ <u>and</u> R_2 <u>of all words</u> $w(c)^{q(c)}$.

7. Groups containing inversions

The preceding discussion goes through with minor modification for G a discontinuous subgroup of $\tilde{H}$. The Dirichlet region Δ is constructed as before, and modified as before to include among its vertices all fixed points on $\partial\Delta$ of elliptic elements of order 2. However, the possibility remains that $t = \Delta \cap \Delta\alpha$ where α of order 2 is an orientation reversing element of G; then, since α maps t to itself and is not an elliptic element reversing t, it must fix all points of t, hence must be inversion in the line ℓ_α containing t. This has the effect that $t\eta = t$, and, conversely, if t is fixed by η, then $t\phi$ is inversion in the line containing t. The corresponding element $t(t\eta)$ of R_1 now gives the relation $t^2 = 1$. The derivation of the presentation of G goes through as before.

Discontinuous subgroups of $\tilde{H}$ are of secondary interest in their own right, but they are useful in the study of Fuchsian groups, since many Fuchsian groups G are naturally obtained as the orientation preserving subgroups $G = \tilde{G}^+$ of a discontinuous subgroup $\tilde{G}$ of $\tilde{H}$, and a fundamental region and presentation for G are easily obtained from those for $\tilde{G}$. We illustrate this in Section 9.

8. Poincaré's Polygon Theorem

We obtained a presentation for a Fuchsian group G (or a discontinuous subgroup of $\tilde{H}$) from very limited information about a fundamental region Δ. The function θ, telling how the sides of Δ fit together in $\partial\Delta$, and the function ω specifying the interior angles of Δ, contain only information which can be read off from a knowledge of Δ. However, the definition of the function η depended on the action of G; explicitly, $t\eta = t\alpha^{-1}$ for the unique α in G such that $t \subseteq \Delta \cap \Delta\alpha$. Poincaré's Polygon Theorem says that if we are given any polygonal region Δ together with a transformation α_t in H (or in $\tilde{H}$) for each side t, such that $t \subseteq \Delta \cap \Delta\alpha_t$, and such that the $c\omega$ are submultiples of 2π, then the group G generated by the α_t is a Fuchsian group (or discontinuous subgroup of $\tilde{H}$) with fundamental region Δ.

THEOREM. <u>Let</u> Δ <u>be a polygonal region and suppose that</u>

(1) <u>for each side</u> t <u>of</u> Δ <u>there is given</u> $\alpha_t \in \tilde{H}$ <u>such that</u> $t \subseteq \Delta \cap \Delta\alpha$.

<u>Define</u> X, θ, ω, η, <u>and</u> σ <u>as before, and assume that</u>

(2) <u>for each cycle</u> c <u>of</u> σ, $c\omega = \frac{2\pi}{q}$ <u>for some integer</u> $q \geqslant 1$.

<u>Then the</u> α_t <u>generate a discontinuous group</u> G <u>with fundamental region</u> Δ.

We will not give a proof. Although the ideas are simple enough, and it is easy to carry out the proof in many special cases, a general argument is a little tricky to formulate without ideas from combinatorial topology. The main idea is that (1) tells us how to match up congruent replicas $\Delta\alpha$ of Δ along the sides of Δ, and (2) ensures that they fit together at the corners. We must show that iteration of this gives us a family of $\Delta\alpha$ that <u>do not overlap</u> and which <u>fill up the hyperbolic plane</u>, in short, a tessellation of H. A common procedure is to construct a purely combinatorial tessellation $\tilde{T}$ with copies of Δ as faces, fitting together at the vertices in the prescribed manner, and then invoke topology to show that the obvious map from $\tilde{T}$ to H is in fact injective.

9. Examples

We illustrate these ideas with a few important examples, including those discussed earlier.

Triangle groups. Let p_1, p_2, p_3 be positive integers such that $\frac{1}{p_1} + \frac{1}{p_2} + \frac{1}{p_3} < 1$. Then there exists a hyperbolic triangle P_1, P_2, P_3 with angles π/p_1, π/p_2, π/p_3. Let $\tilde{\Delta}$ be the region bounded by this triangle, with sides t_1, t_2, t_3 opposite P_1, P_2, P_3. Let ρ_1, ρ_2, ρ_3 be inversion in the lines containing these three sides. Then $t_i = \tilde{\Delta} \cap \tilde{\Delta}\rho_i$. The cycles of $\tilde{\Delta}$ are $c = (t_2,t_3)$, (t_3,t_1), (t_1,t_2), and $\omega = 2\pi/q$ for $q = p_1$, p_2, p_3. Thus, by Poincaré's Theorem, $\tilde{\Delta}$ is a fundamental region for a discontinuous subgroup $\tilde{G}$ of $\tilde{H}$ with a presentation

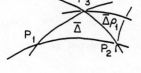

$$\tilde{G} = \langle \rho_1, \rho_2, \rho_3 : \rho_1^2 = \rho_2^2 = \rho_3^2 = (\rho_2\rho_3)^{p_1} = (\rho_3\rho_1)^{p_2} = (\rho_1\rho_2)^{p_3} = 1 \rangle.$$

Let G be the Fuchsian group $G = \tilde{G}^+$, generated by $\alpha_1 = \rho_2\rho_3$, $\alpha_2 = \rho_3\rho_1$, and $\alpha_3 = \rho_1\rho_2$. Then G has a fundamental region $\Delta = \tilde{\Delta} \cup \tilde{\Delta}\rho_1$. Let Δ as shown, have sides $t_1 = \Delta \cap \Delta\alpha_3$, $t_1 = \Delta \cap \Delta\alpha_3^{-1}$, $t_2 = \Delta \cap \Delta\alpha_2$, $t_2 = \Delta \cap \Delta\alpha_2^{-1}$. Then the vertex relations are as follows:

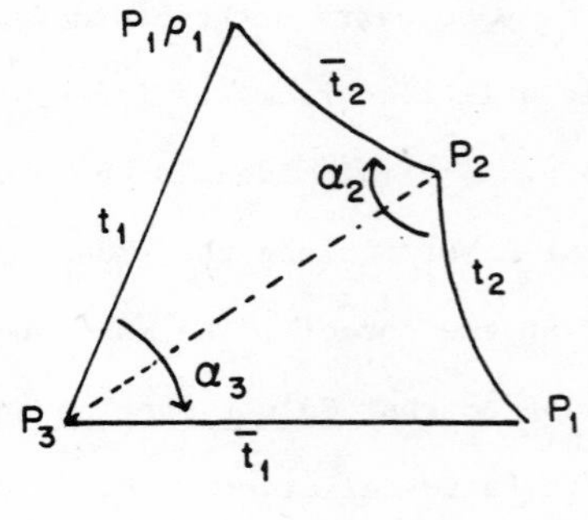

at P_1, $(\alpha_3\bar{\alpha}_2)^{p_1} = 1$;

at P_2, $\alpha_2^{p_2} = 1$;

at $P_1\rho_1$, $(\alpha_2\bar{\alpha}_3)^{p_1} = 1$;

at P_3, $\alpha_3^{p_3} = 1$.

As expected, the relations at P_1 and $P_1\rho_1$ are equivalent. Dropping one of them we obtain directly by Poincaré's method the presentation $G = \langle \alpha_2, \alpha_3 : (\alpha_3\bar{\alpha}_2)^{p_1} = 1, \alpha_2^{p_2} = 1, \alpha_3^{p_3} = 1 \rangle$, which differs only notationally from that obtained before.

<u>Triangle groups with vertices at infinity</u>. Suppose now that the vertex P_3 is at infinity, where the sides t_2 and t_3 now meet at angle 0. Then $\alpha_3 = \rho_1\rho_2$ is a parabolic element of infinite order, and, in accordance with our theorems, there is no relation $\alpha_3^{p_3} = 1$ with $p_3 \neq 0$. We obtain a presentation for $\tilde{G}$ as before, with the relation $(\rho_1\rho_2)^{p_3} = 1$ deleted. For the Fuchsian group G we obtain a presentation $G = < \alpha_1, \alpha_2, \alpha_3 : \alpha_1^{p_1} = \alpha_2^{p_2} = \alpha_1\alpha_2\alpha_3 = 1 >$. Eliminating α_3 by a Tietze transformation gives a presentation $G = < \alpha_1, \alpha_2 : \alpha_1^{p_1} = \alpha_2^{p_2} = 1 >$, whence $G \simeq G_{p_1} \star G_{p_2}$, the free product of two cyclic groups of orders p_1 and p_2.

If two vertices, say P_2 and P_3 are at infinity, we obtain now $G = < \alpha_1, \alpha_2 : \alpha_1^{p_1} = 1 > \simeq G_{p_1} \star G_\infty$, the free product of a cyclic group of order p_1 with an infinite cyclic group.

If all three vertices are at infinity, we obtain

$$\tilde{G} = < \rho_1, \rho_2, \rho_3 : \rho_1^2 = \rho_2^2 = \rho_3^2 = 1 > \simeq C_2 \star C_2 \star C_2,$$

and

$$G = < \alpha_1, \alpha_2 : \emptyset > \simeq C_\infty \star C_\infty, \text{ a free group of rank 2.}$$

<u>Schottky groups</u>. Let $\tilde{\Delta}$ be a region in H bounded by three lines ℓ_1, ℓ_2, ℓ_3 that do not intersect in H. If each pair intersects in a point on ∂H, we have the case treated above of a triangle with all three vertices at infinity, but in any case we have no vertex relations, whence the groups $\tilde{G}$ and G have the same presentations as for the case of the triangle in particular, G is a free group of rank 2.

More generally, let $\tilde{\Delta}$ be bounded by n lines $\ell_1,\ldots,\ell_n$ that do not intersect in H; we may suppose that they are the parts in H of n circles orthogonal to <u>R</u> with disjoint interiors. Then exactly the same argument shows that $\tilde{G}$ is the free product of n groups of order 2 and that G is free of rank n-1.

The modular group. The standard fundamental region Δ for this group G, as described earlier, is bounded by a triangle with one vertex at infinity. The argument above, or, as we have seen, a direct application of Poincaré's Theorem, shows that Δ is the fundamental region for $G = \langle \alpha, \beta : \alpha^2 = \beta^3 = 1 \rangle$, where α is elliptic of order 2 with fixed point i and β elliptic of order 3 with fixed point $\omega = \frac{1 + i\sqrt{3}}{2}$.

This group arises in various contexts, in algebra, analysis, and number theory. We indicate a basic way in which it arises in algebra.

Let A be the free abelian group of rank 2, $A = \langle a_1, a_2 : a_1a_2 = a_2a_1 \rangle$. As is common with abelian groups, we write the group operation as addition instead of multiplication, thus $A = \langle a_1, a_2 : a_1 + a_2 = a_2 + a_1 \rangle$. One can think of A as the additive group generated by two linearly independent vectors a_1 and a_2 in the plane. Now, as in linear algebra, a mapping α from A into A is defined by its action on the generators a_1 and a_2, and hence by a matrix $M = \begin{pmatrix} a, & b \\ c & d \end{pmatrix}$ with entries a, b, c, $d \in \underline{Z}$. Now α has an inverse of the same form just in case its determinant is an element of $\underline{Z}$ with a reciprocal in $\underline{Z}$, that is, $\det M = ad - bc = \pm 1$. Thus the group of all automorphisms of A can be identified with the group of all matrices $\begin{pmatrix} a & b \\ c & d \end{pmatrix}$, $ad - bc = \pm 1$, which we denote by $GL(2,\underline{Z})$. As with matrices over a field, we define $SL(2,\underline{Z})$ to be the orientation preserving subgroup of all such matrices M with $\det M = +1$.

Since $\underline{Z}$ is a subring of $\underline{R}$, we can identify SL(2,Z) with a subgroup of $SL(2,\underline{R})$, and hence $PSL(2,\underline{Z})$ with a subgroup of $PSL(2,\underline{R}) = \mathcal{H}$. Thus $PSL(2,\underline{Z})$ is the quotient of $SL(2,\mathbb{Z})$ by its center, consisting of the two matrices I and -I.

THEOREM. $PSL(2,\underline{Z})$, as a subgroup of $\mathcal{H}$, is the modular group.

Proof. The generators $\alpha : z \mapsto -1/z$ and $\gamma = \alpha\beta : z \mapsto z + 1$ are given by the matrices $A = \begin{pmatrix} 0 & -1 \\ 1 & 0 \end{pmatrix}$ and $C = \begin{pmatrix} 1 & 1 \\ 1 & 0 \end{pmatrix}$. It will suffice to show that $SL(2,\underline{Z})$ is generated by A, C, and $A^2 = -1 = \begin{pmatrix} -1 & 0 \\ 0 & -1 \end{pmatrix}$. For this, let $M = \begin{pmatrix} a & b \\ c & d \end{pmatrix}$, a, b, c, $d \in \underline{Z}$, det M = +1.

We argue by induction on m(M), the minimum of $|a|$, $|c|$. We note that, replacing M by $M' = AM = \begin{pmatrix} -c & d \\ a & b \end{pmatrix}$, we may always suppose that $|c| \leqslant |a|$. If $|c| = 0$, then $c = 0$, det M = ad = 1, whence $a = d = \pm 1$, $M = \pm\begin{pmatrix} 1 & \pm b \\ 0 & 1 \end{pmatrix} = \pm C^{\pm b}$, and we are done.

Let $0 < |c| \leqslant |a|$. Then, for some $q \in \underline{Z}$, $a = qc + c'$ where $|c'| < |c|$. Now $M' = C^{-q}M = \begin{pmatrix} 1 & -q \\ 0 & 1 \end{pmatrix}\begin{pmatrix} a & b \\ c & d \end{pmatrix} = \begin{pmatrix} a-qc & b-qd \\ c & d \end{pmatrix} = \begin{pmatrix} c' & b-qd \\ c & d \end{pmatrix}$, with $m(M') = |c'| < |c|$, and, by induction, we are done. □

Regular tessellations. A regular polygonal region Π is a region in H bounded by a finite simple polygon with all sides congruent and all angles equal. A regular tessellation T of type (p,q) is one in which the tiles are congruent regular p-gonal regions with q meeting at each vertex. We saw earlier that tessellations of the sphere of type (p,q) exist just in case $\frac{1}{p} + \frac{1}{q} > \frac{1}{2}$, and tessellations of the Euclidean plane of type (p,q) exist just in case $\frac{1}{p} + \frac{1}{q} = \frac{1}{2}$. We will now show that a tessellation T of the hyperbolic plane H of type (p,q) exists if (and only if) $\frac{1}{p} + \frac{1}{q} < \frac{1}{2}$.

If $\frac{1}{p} + \frac{1}{q} < \frac{1}{2}$, there exists a triangular region Δ in H with angles π/p, π/q, $\pi/2$. Let $\tilde{G}$ be the subgroup of $\tilde{H}$ generated by inversions ρ_1, ρ_2, ρ_3 in the lines ℓ_1, ℓ_2, ℓ_3 containing the sides of Δ opposite the vertices with these angles, and let T be the resulting tessellation of H. At the vertex P_1 of Δ with angle π/p there will be 2p tiles $\Delta\alpha$ about P_1 (indeed for $\alpha = 1, \rho_1, \rho_1\rho_2, \rho_1\rho_2\rho_1, \ldots, (\rho_1\rho_2)^{p-1}\rho_1$) forming a regular p-gonal region Π with angles $2\pi/q$. The set of all $\Pi\alpha$ for α in G evidently forms a tessellation Z of type (p,q).

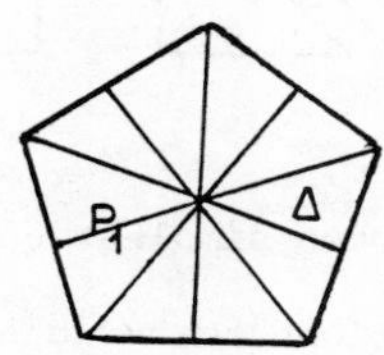

The group $\tilde{G}$ with fundamental region Δ is evidently the full symmetry group of Θ. The Fuchsian subgroup $G = \tilde{G}^+$ of orientation preserving symmetries of Θ is the triangle group $G = < \alpha_1, \alpha_2 : \alpha_1^p = \alpha_2^q = (\alpha_1\alpha_2)^2 = 1 >$. There need not exist any Fuchsian group with fundamental region Π (for example, if p is odd), but we shall encounter such groups below in the case that $p = q = 4g$, $g \geqslant 2$.

10. Surface groups

To avoid entering into technicalities from topology, we define a (compact orientable) surface Σ to be a nonoverlapping union of a finite number of oriented cells K, homeomorphic to triangles, in such a way that each side appearing on one cell K appears as a side of exactly one other cell K', with opposite orientation.

The simplest example is the sphere $\Sigma_0 = S$, decomposed into four (spherical) triangles by projection of an inscribed tetrahedron. The next simplest is the torus (anchor ring) Σ_1, obtainable from two triangles by identifying corresponding sides a, $\bar{a}$, b, $\bar{b}$, c, $\bar{c}$ as shown in the figure.

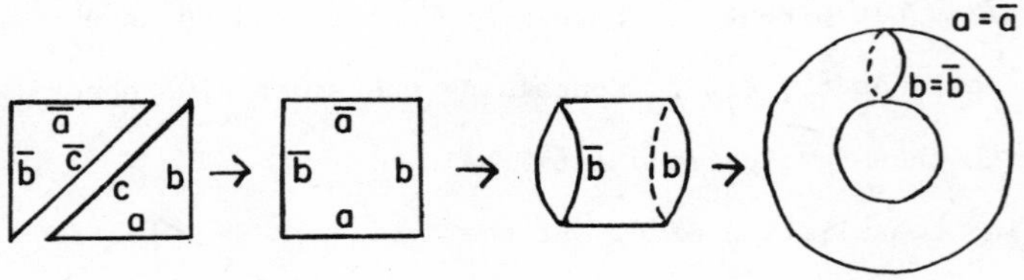

These two simplest cases are 'spherical' and 'Euclidean' and are thus exceptions to our discussion of the hyperbolic case. It is an elementary fact, which we shall illustrate in the smallest case, that every other (compact orientable) surface Σ can be obtained from the fundamental region Δ for a Fuchsian group G by identifying pairs of sides t and $\bar{t} = t\eta$.

Define the <u>quotient space</u> H/G of a Fuchsian group G to be the set of <u>orbits</u> $QG = \{Q\alpha : \alpha \in G\}$ of points $Q \in H$. The projection $\Pi : H$ onto H/G, carrying each Q to QG, is evidently bijective from Δ onto H/G except on $\partial\Delta$, where $t\pi = \bar{t}\pi$. Thus $H/G = \Delta/G$, and the quotient space H/G can be obtained as the surface $\Sigma = \Delta/G$ resulting from Δ by identifying the sides in pairs t, $\bar{t}$, according to α_t.

We will not define the <u>fundamental group</u> $\pi_1(\Sigma)$ of a surface Σ, but note only the easy but important fact that, for Σ obtained as above, $\pi_1(\Sigma) \simeq G$. (For the sphere Σ_0, $\pi_1(\Sigma_0) = 1$, and for the torus Σ_1, $\pi_1(\Sigma_1)$ is free abelian of rank 2.)

It is easy to see that $\pi : H \to \Sigma$ is continuous and indeed a <u>local homeomorphism</u>: every point Q of H is contained in a disc D such that the restriction of π to D is a homeomorphism. In the case at hand, where G contains no elliptic elements, the analytic structures on the $D\pi$, inherited from those on the $D \subseteq H$, 'match up' in the sense that, on their intersections, the passage from one to the other is analytic, and Σ becomes an <u>analytic</u> (indeed, <u>hyperbolic</u>) <u>surface</u> (<u>manifold</u> of dimension 2). (In the more general case, where G contains an elliptic element of order q fixing a vertex V of Δ, the map on a small disc D at V acts like the map $z \mapsto z^q$ on the unit disc, and Σ is a <u>Riemann surface</u> whose analytic structure has <u>branch points</u> at such V.)

To obtain surfaces Σ as Δ/G, it turns out that we can manage under certain simplifying assumptions on Δ and G. The first is that G contains no elliptic elements, and the second that all the vertices V of Δ belong to a single orbit, hence have the same image in Σ. This implies that R_2 contains only a single element r (apart from redundant cyclic permutations of r and r^{-1}), containing each x in X exactly once. More precisely, σ has a single orbit c, and r is the product, starting at some element, of the x_i in this orbit, in order. To obtain a more economical presentation for G we choose a subset X_0 of X containing exactly one from each pair t, $\bar{t}$, and, by Tietze transformations, use the relations $t\bar{t} = 1$ to eliminate the $\bar{t}$ not in X_0, at the same time expressing r in terms of X_0 by replacing each $\bar{t}$ by t^{-1} for $t \in X_0$. We obtain thus a presentation $G = \langle X_0 : r = 1 \rangle$ with a <u>single defining relation</u>.

We have yet to construct Δ, and, with it, G. Since η is an involution on the set of sides, without fixed element, the number of sides must be even, n = 2k for some k. Then θ consists of a single cycle of even length 2k, whence θ is an <u>odd</u> permutation. Also, we have stipulated that σ must consist of a single cycle, of length 2k, whence σ is an <u>odd</u> permutation. It follows from σ = θη that η must be an <u>even</u> permutation; since η consists of k transpositions $(t,\bar{t})$, this requires that k be even, k = 2g, and n = 4g.

Since σ has a single cycle c, and G contains no elliptic element, we must have $c\omega = 2\pi/q(c)$ with $q(c) = 1$, that is, the sum of the interior angles is $\Sigma\theta_t = 2\pi$. The area of Δ is now $A(\Delta) = \Sigma(\pi - \theta_t) - 2\pi = 4g\pi - \Sigma\theta_t - 2\pi = 4g\pi - 2\pi - 2\pi = 4(g - 1)\pi$. Since Δ is compact, $A(\Delta) > 0$, and thus $g \geqslant 2$.

Let $g \geqslant 2$ and let Δ be a 4g-gonal region. It remains to choose the interior angles, $t\omega$, subject to $\Sigma t\omega = 2\pi$, and to choose the involution η. It turns out not to matter how we choose the $t\omega$, and thus we may as well choose them all equal, $t\omega = 2\pi/4g$; thus, we take Δ to be a regular 4g-gonal region. Since c has length 4g, there are 4g tiles Δα at each vertex, and the tessellation T is regular of type (4g,4g).

From here on, to avoid notational complexity, we treat only the smallest, but entirely typical, case, that the <u>genus</u> g = 2. Now Δ is a regular octagonal region, and it remains to choose η, that is, to decide the cyclic order θ of the eight edges a, $\bar{a}$, b, $\bar{b}$, c, $\bar{c}$, d, $\bar{d}$. Here η is subject only to the constraint that σ, defined by the equation $\sigma = \theta\eta$, must be a single cycle of length 8. The resulting surface Σ is the same for all η satisfying this constraint. Perhaps the most natural choice of η is to pair opposite sides, that is to take $\theta = (a,b,c,d,\bar{a},\bar{b},\bar{c},\bar{d})$; this gives $\sigma = (a,\bar{b},c,\bar{d},\bar{a},b,\bar{c},d)$, and $G = \langle a,b,c,d : a\bar{b}c\bar{d}\bar{a}b\bar{c}d = 1 \rangle$, or, changing notation $G = \langle a,b,c,d : abcda^{-1}b^{-1}c^{-1}d^{-1} = 1 \rangle$. A second choice is $\theta = (a, b, \bar{a}, \bar{b}, c, c, \bar{c}, \bar{d})$, giving $\sigma = (a, \bar{b}, \bar{c}, d, c, \bar{d}, \bar{a}, b)$. After a cyclic permutation and a relettering, this yields a presentation $G = \langle a, b, c, d : a\,b\,a^{-1}\,b^{-1}\,c\,d\,c^{-1}\,d^{-1} = 1 \rangle$

It is an elementary but not entirely trivial exercise to show by Tietze transformations that these two presentations define isomorphic groups; the Tietze transformations used can indeed be paralleled by transformations of Δ by 'cutting and pasting': cutting Δ into two parts, Δ_1 containing a side t and Δ_2 the side $\bar{t}$, and forming $\Delta' = \Delta_1 \cup \Delta_2\alpha^{-1}$ by identifying t with $\bar{t} = t\alpha^{-1}$.

We end this excursion to the borders of topology by sketching a proof that the surface Σ_2, obtained as above for genus g = 2, is a 'sphere with two handles', or (homeomorphic to) the surface of a solid figure eight, made of two solid rings ∞ welded together.

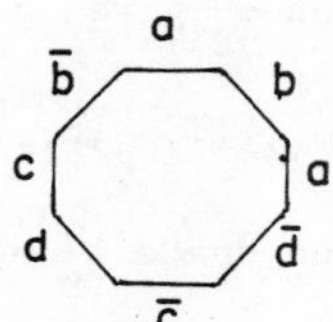

For this, we take $\theta = (a,b,\bar{a},\bar{b},c,d,\bar{c},\bar{d})$ and label the sides of the octagonal region Δ accordingly. Before identifying any sides we take a step backward, cutting into two pieces, Δ_1 and Δ_2, each with a side f

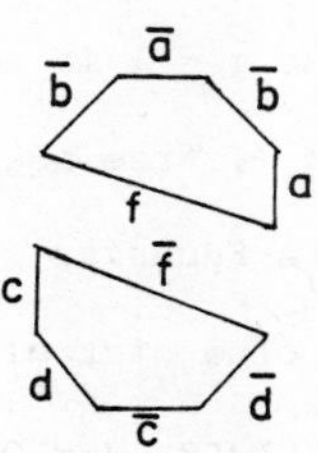

or $\bar{f}$, to be reunited later. Now we identify sides a and $\bar{a}$, b and $\bar{b}$ of Δ_1; this is the same as identifying the opposite sides of a square to obtain a torus, except that now we obtain a torus $\bar{\Delta}_1$ with a hole, whose boundary is labelled by f. Similarly, identification of the sides c, $\bar{c}$ and d, $\bar{d}$ of Δ_2 yields a second torus $\bar{\Delta}_2$ with a hole, now with boundary labelled by $\bar{f}$. The last step is to unite $\bar{\Delta}_1$ and $\bar{\Delta}_2$ by identifying f and $\bar{f}$. The result is the surface of a rather primitive 'pretzel with two holes' shown in the figure.

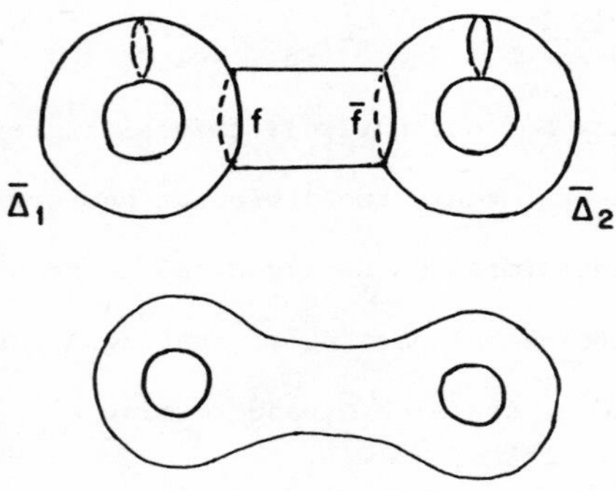

For all $g \geqslant 2$ a similar construction using a regular 4g-gon with interior angles $\frac{2\pi}{4g}$ gives a group G with a presentation equivalent to

$$G = \langle a_1,\ldots,a_g,b_1,\ldots,b_g : a_1b_1a_1^{-1}b_1^{-1}\ldots a_gb_ga_g^{-1}b_g^{-1} = 1 \rangle,$$

with quotient space $\Sigma_g = H/G$ homeomorphic to the surface of a pretzel with g holes.

The case $g = 1$ is an exception only in that the group $G = \langle a_1, b_1 : a_1b_1\bar{a}_1\bar{b}_1 = 1 \rangle$, free abelian of rank 2, is a Euclidean group acting on E rather than a Fuchsian group; the quotient surface Σ_1 is a torus, that is, the surface of a pretzel with only one hole. The case $g = 0$ can be viewed as the degenerate case of the discontinuous group $G = 1$ acting on the sphere S, with quotient space $S/G \simeq S$ a pretzel with no holes.

11. Classification of Fuchsian groups

We sketch a proof of the following.

THEOREM. Every Fuchsian group G with compact fundamental region has a presentation of the form

$$(*)\quad G = \langle x_1,\ldots,x_k,y_1,\ldots,y_g,z_1,\ldots,z_g :$$
$$x_1^{q_1} = \ldots = x_k^{q_k} = x_1\ldots x_k y_1 z_1 y_1^{-1} z_1^{-1} \ldots y_g z_g y_g^{-1} z_g^{-1} = 1 \rangle,$$

for $k, g \geqslant 0$, all $q_i \geqslant 2$, and, if $g = 0$, $k \geqslant 3$. Moreover, apart from permuting the q_i, two groups with different presentations of this form are not isomorphic.

Outline of proof. Let Δ be compact. Then every side $t \in S$ ends at some vertex V of Δ in H, whence t appears exactly once in s_c for some relator $r_c = s_c^{q(c)} = (t_1\ldots t_n)^{q(c)}$ for $c = (t_1,\ldots,t_n)$ a cycle of σ. Moreover, since θ is transitive on the set S of sides, no subset R' of R_2 except the empty set and R_2 itself has the property that if t occurs in some r_c in R' then $\bar{t}$ occurs in some $r_{c'}$ in R'.

From now on we use the relators R_1 freely, identifying $\bar{t}$ with t^{-1}.

Suppose that $|R_2| > 1$. If c_1 is any cycle of σ, it follows that s_{c_1} contains some t_1 but not $\bar{t}_1$, and contains t_1 only once. After a succession of Tietze transformations we can replace the generator t_1 by s_{c_1}, and thus the relator r_{c_1} by $t_1^{q_1}$ where $q_1 = q(c_1)$. Let R_2' be the remaining set of relators. One sees that, if $|R_2'| > 1$, then any s_{c_2} in R_2' contains some $t_2 \neq t_1^{\pm 1}$ but not $\bar{t}_2$, and, by Tietze transformations one can replace r_{c_2} by $t_2^{q_2}$.

We can continue thus until we arrive at a presentation of the form $G = \langle t_1,\ldots,t_{k-1},u_1,\ldots,u_m : t_1^{q_1} = \ldots = t_{k-1}^{q_{k-1}} = w^{q_k} = 1 \rangle$, where we may now suppose that w contains all t_i but no t_i^{-1}, and all u_j and u_j^{-1}, exactly once each. We introduce a new generator t_k with a defining relator $t_k w$, and then replace the relator $w^{q_k} = 1$ by equivalent $t_k^{q_k} = 1$. This gives $G = < t_1,\ldots,t_k,u_1,\ldots,u_m : t_1^{q_1} = \ldots = t_k^{q_k} = w' = 1 >$ where $w' = t_k w$.

It is an exercise in Tietze transformations to show that, by changing the generators u_i, we can rewrite w' in the form $w' = t_1 \ldots t_k q$ where q contains only the u_i and u_i^{-1}, each exactly once. It is a further and more difficult exercise to show that further Tietze transformations on the u_i transforms q to the form $y_1 z_1 y_1^{-1} z_1^{-1} \ldots y_g z_g y_g^{-1} z_g^{-1}$ where $2g \leqslant m$. An appeal to the transitivity of θ now enables us to prove that $2g = m$.

The uniqueness of this presentation is routine. The q_i are the common orders of the members of the conjugacy classes of maximal finite cyclic subgroups of G. The genus g is the rank of the free abelian group obtained by first abelianizing G and then dividing by the subgroup of all elements of finite order.

We note that if $g = 0$ and $k = 0$, then $G = 1$. If $g = 0$ and $k = 1$, then G is cyclic of order q_1, not a Fuchsian group with compact fundamental region Δ (since q_1 images of compact Δ cannot fill out H). If $g = 0$ and $k = 2$, then G is cyclic of order the greatest common divisor of q_1 and q_2.

The same methods as those used above show that if G is finitely generated, that is, if Δ has finitely many sides, but if now Δ has vertices on $\underline{R}^*$ (which correspond to no relation in R_2), then G is a free product of cyclic groups.

12. The Riemann-Hurwitz formula

We again assume that Δ has finitely many sides, possibly with some vertices on $\underline{R}^*$, but with no side on $\underline{R}^*$, whence Δ has finite area $A(\Delta)$.

If Δ has $S = 2n$ sides, then Δ has $2n$ vertices V. If $V \in \underline{R}^*$, then the interior angle at V is 0. If $c = (t_1,\ldots,t_n)$ is a cycle of σ, and $V_1,\ldots,V_n$ are the ends of $t_1,\ldots,t_n$ (in the positive sense on $\partial\Delta$), then these $V_i \in H$, and every vertex in H occurs just once in one such cycle. Now, we have seen, the sum of the interior angles at $V_1,\ldots,V_n$ is $\frac{2\pi}{q(c)}$, whence the sum of all the interior angles at vertices of Δ is $2\pi\cdot\Sigma\,\frac{1}{q_i}$, summed over $q_i = q(c_i)$ for all k cycles c_i of σ. It follows that $\kappa(\partial\Delta) = \pi\cdot 2n - 2\pi\cdot\Sigma\,\frac{1}{q_i} = 2\pi\left(n - \Sigma\,\frac{1}{q_i}\right)$.

We now pass to the presentation (*) (or the corresponding presentation of G as a free product of cyclic groups). We started with n generators consisting of one out of each pair t, $\bar{t}$ of the $2n$ sides, and adjoined one more generator t_k by a Tietze transformation. Thus in the presentation (*) there are $n + 1 = k + 2g$ generators. Hence we can rewrite the equation above in the form

$$\kappa(\partial\Delta) = 2\pi\left(2g - 1 - \Sigma\left(1 - \frac{1}{q_i}\right)\right).$$

It follows that

$$A(\Delta) = 2\pi\left(2g - 2 - \Sigma\left(1 - \frac{1}{q_i}\right)\right).$$

It is not difficult to see that if G_1 and G_2 are two Fuchsian groups with fundamental regions Δ_1 and Δ_2 of finite area, and if G_2 is a subgroup of G_1 of finite index j, then $A(\Delta_2) = j\cdot A(\Delta_1)$. In particular, if $G_1 = G_2$, whence $j = 1$, we have $A(\Delta_1) = A(\Delta_2)$; thus $A(\Delta)$ depends only on G and not on the choice of fundamental region. Thus we may define

$$\mu(G) = 2g - 2 - \Sigma\left(1 - \frac{1}{q_i}\right).$$

The assertion above can now be stated more concisely.

THEOREM (Riemann-Hurwitz Formula). _Let_ G_1 _and_ G_2 _be two Fuchsian groups with fundamental regions of finite area, and let_ G_2 _be a subgroup_ *of finite index* j *in* G_1. *Then*

$$j = \frac{\mu(G_2)}{\mu(G_1)} .$$

Notes and references

1. A number of tessellations of H by fundamental regions of Fuchsian groups (reproduced from Klein and Fricke) are shown, with commentary, in the book of Magnus. A very elementary and geometric discussion of some Fuchsian groups, based on work of Dyck, is given in Chapter 18 of Burnside.

2. The theorem in Section 3 gives half the proof of a theorem of Fenchel: Every finitely generated Fuchsian group G contains a normal subgroup N of finite index that is torsion free (without nontrivial elements of finite order). The proof reduces easily to the case that G is a triangle group of the form $G = \langle \sigma_1, \sigma_2 : \sigma_1^a = \sigma_2^b = (\sigma_1\sigma_2)^c = 1 \rangle$, where $\frac{1}{a} + \frac{1}{b} + \frac{1}{c} < \frac{1}{2}$. Now, by the theorem in Section 3, if we can find a map ϕ from G onto a finite group F under which the images $\sigma_1\phi$, $\sigma_2\phi$, and $(\sigma_1\sigma_2)\phi$ retain orders exactly a, b, c, then the kernel N of ϕ will be a torsion free subgroup of finite index $|F|$.

The existence of such ϕ amounts to the existence of a finite group containing elements x, y such that x, y, and xy have orders a, b, c. This was proved by G. A. Miller in 1900 for all $a, b, c \geqslant 2$. Miller's result was not known to Fenchel, who stated it as a conjecture, or to Fox, who gave a proof. For references see Brenner-Lyndon, Lyndon-Schupp.

3. Cayley introduced the 'colour graph' of a group G, relative to a set X of generators. This graph has vertices all elements of G, with a (directed) edge with 'color' $x \in X$ from g_1 to g_2 just in case $g_1x = g_2$.

The same graph was introduced by Dehn as the Gruppenbild. See Coxeter-Moser, Section 3.3 and following.

4. Poincaré's paper in which he defines Fuchsian groups and obtains presentations for them is still very readable.

5. The function $\sigma = \theta\eta$ was used formally by Hoare-Karrass-Solitar in a broader context, where it takes the form of a symmetric function, represented by a graph with undirected edges, called the coinitial graph or the star graph of a presentation $\langle X:R \rangle$. The vertices are all x and x^{-1} for $x \in X$. There is an edge between two vertices x and y for each cyclic permutation of r or r^{-1} for some $r \in R$ that begins with xy^{-1}. The same function had entered incidentally in the proofs of Whitehead's theorem on automorphisms of free groups. For references, see Lyndon-Schupp.

6. A proof of Poincaré's Polygon Theorem was first offered by Poincaré, and later by others, in various degrees of generality. See, in addition to Poincaré, Maskit, de Rham, and Beardon.

7. For surface groups and the relevant topological concepts, see Massey. We mention now compact nonorientable surfaces. Such a surface is the nonoverlapping union of a finite number of 'triangles', with each edge appearing on exactly two triangles, but now in such a way that the triangles cannot be oriented to make each edge occur with opposite orientation on the two triangles.

The simplest example of this is the real projective plane E^*. From our construction of E^* as the image of a sphere S under stereographic projection, we conclude that E^* is homeomorphic to the result of identifying antipodal points of S, or, more simply, of identifying antipodal points on the boundary of a hemisphere Δ of S. We divide the boundary circle of Δ into four equal parts, and then deform Δ into a square with these parts as sides. In the figure, the pairs of opposite sides must be identified in such a way that the directions indicated by the arrows match up. If we suppose this identification made, and divide Δ into two triangles by a diagonal, it is easy to see that the two triangles cannot be oriented in such a way that each edge occurs on one triangle with positive orientation and on the other with negative orientation.

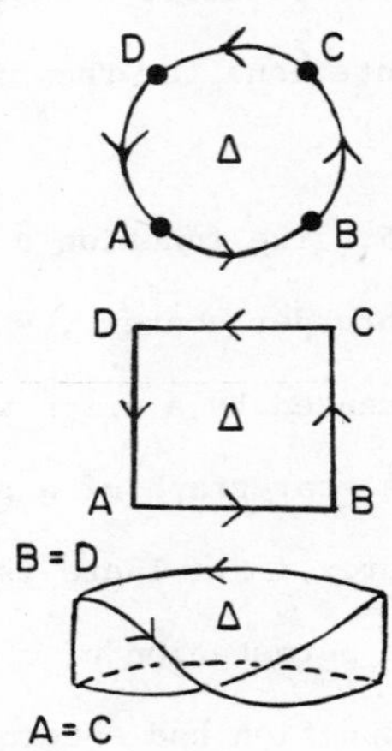

In fact, it is possible in 3-space to identify one pair of sides, say BC and DA, (thus identifying B = D and C = A) to obtain a twisted Möbius strip, with boundary a simple closed curve in 3-space. But it is not possible in 3-space to identify the remaining pair of sides.

8. Hoare-Karrass-Solitar proved (purely group theoretically without use of geometry or analysis) that if G_1 is a Fuchsian group with a fundamental region of finite area and G_2 is any subgroup of G_1, then either G_2 has finite index in G_1 and is another such subgroup, or G_2 has infinite index and is a free product of cyclic groups.

9. Discontinuous subgroups of $\tilde{H}$ are often called <u>noneuclidean crystallographic</u> (NEC) groups: see Macbeath.

10. For a general reference on Fuchsian groups we again recommend Beardon. For something on their history and their connections with differential equations, analysis, and number theory, we recommend selective reading in Lehner.

Problems

Problem 1. Find a Dirichlet region for the cyclic group G generated by α, where α is elliptic of finite order, parabolic, or hyperbolic.

Problem 2. (a) Let G in $\tilde{H}$ be generated by reflections in the three sides of a triangle. For arbitrary P inside the triangle, what is the Dirichlet region for G?

(b) Let G in $\tilde{H}$ be generated by reflection in the three (half) circles with radius 1 and centers at (-5,0), (0,0), (+5,0). For conveniently chosen P, find the Dirichlet region for G.

Problem 3. (a) For G as in Problem 1, (a) and (b), find the corresponding Cayley tessellation.

(b) For the usual fundamental region for the modular group, as shown in the Example of Section 5, show a representative piece of the Cayley tessellation.

Problem 4. (a) What are the possible Fuchsian groups having a Dirichlet region with four sides?

(b) What are the possible Fuchsian groups with a six-sided fundamental region and $\theta = (a, b, c, \bar{a}, \bar{b}, \bar{c})$?

<u>Problem 5</u>. (a) The fundamental group of the complement in E^3 of the trefoil knot (with ends spliced together) has a natural presentation $G = \langle a, b : aba = bab \rangle$. Show that it also has a presentation $G = \langle c, d : c^2 = d^3 \rangle$, and that the quotient of G by its center is isomorphic to the modular group.

(b) Show that $G = SL(2,\underline{Z})$ has a presentation of the form $G = \langle c, d : c^2 = d^3, c^4 = 1 \rangle$.

<u>Problem 6</u>. (a) Find a fundamental polygon for a group G with presentation $G = \langle x, y, z : x^3 = xyzy^{-1}z^{-1} \rangle$.

(b) The same for $G = \langle x_1, x_2, y, z : x_1^3 = x_2^5 = x_1x_2yzy^{-1}z^{-1} \rangle$.

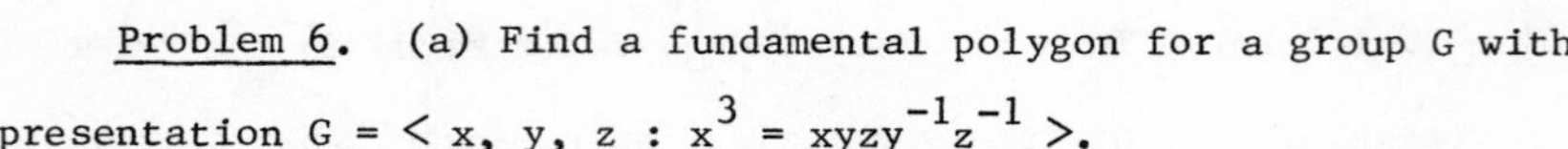

<u>Problem 7</u>. Supply the details of the proof of the theorem in Section 11 in the case that k = 2 and g = 1 or g = 2.

<u>Problem 8</u>. (a) If Δ is a fundamental region for a Fuchsian group G such that $H/G = \Sigma_g$ is a compact orientable surface of genus g, what is the area of Δ ?

(b) What is the smallest possible area of a compact fundamental region for a Fuchsian group?

<u>Problem 9</u>. (a) Find a torsion free normal subgroup of finite index in the modular group $G = \langle a, b : a^2 = b^3 = 1 \rangle$.

(b) The same for $G = \langle a, b : a^2 = b^3 = (ab)^7 = 1 \rangle$.

<u>Problem 10</u>. (a) Construct the Cayley graph for the presentation $G = \langle x, y : x^2 = y^n = (xy)^2 = 1 \rangle$ of the dihedral group D_{2n}.
(b) The same for the groups $G = \langle x, y : x^2 = y^3 = (xy)^r = 1 \rangle$ for $r = 3, 4, 5, 6$, and 7.

REFERENCES

Albert, A.A. & Sandler, R. (1968). An Introduction to Finite Projective Plane. Holt, Rinehart & Winston.

Artin, E. (1957). Geometric Algebra, Interscience.

Beardon, A.F. (1983). The Geometry of Discrete Groups, Springer.

Brenner, J.L. & Lyndon, R.C. (1984). A theorem of G. A. Miller on the order of the product of two permutations. I. Jnānābha 14, 1-16.

Coxeter, H.S.M. (1961). Introduction to Geometry, Wiley.

Coxeter, H.S.M. (1961). Non-Euclidean Geometry, 4th edition, Univ. of Toronto.

Coxeter, H.S.M. (1968). Twelve Geometric Essays, Southern Illinois Univ. Press.

Coxeter, H.S.M. (1973). Regular Polytopes, reprint of 2nd edition, Dover.

Coxeter, H.S.M. & Moser, W.O. (1965). Generators and Relations for Discrete Groups, 2nd edition, Springer.

de Rham, G. (1971). Sur les polygones générateurs de groupes fuchsiens. L'Enseignment Math. 17, 49-61.

Ford, L.R. (1951). Automorphic Functions, 2nd edition, Chelsea.

Guggenheimer, H.W. (1967). Plane Geometry and Its Groups, Holden-Day.

Hall, M. (1959). The Theory of Groups, Macmillan.

Hoare, A.H.M., Karrass, A. & Solitar, D. (1971). Subgroups of finite index of Fuchsian groups, Math. Zeit. 120, 289-298.

Hoare, A.H.M., Karrass, A. & Solitar, D. (1972). Subgroups of infinite index in Fuchsian groups, Math. Zeit. 125, 59-69.

Hoare, A.H.M., Karrass, A. & Solitar, D. (1973). Subgroups of NEC groups, Comm. Pure Appl. Math. 26, 731-744.

Johnson, D.L. (1980). Topics in the Theory of Group Presentations. Cambridge Univ. Press.

Lehner, J. (1964). Discontinuous Groups and Automorphic Functions. Amer. Math. Soc.

Lyndon, R.C. & Schupp, P.E. (1977). Combinatorial Group Theory, Springer.

Macbeath, A.M. (1967). The classification of non-Euclidean plane crystallographic groups. Canad. J. Math. 6, 1192-1205.

Macbeath, A.M. (1976). Groups of hyperbolic crystallography. Math. Proc. Camb. Phil. Soc. 79, 235-249.

Magnus, W. (1974). Noneuclidean Tessellations and their Groups. Academic Press.

Magnus, W., Karrass, A. & Solitar, D. (1966). Combinatorial Group Theory. Wiley.

Manning, H.P. (1914). Geometry of Four Dimensions. Macmillan.

Maskit, B. (1971). On Poincaré's theorem for fundamental polygons. Adv. in Math. 7, 219-230.

Miller, C.F. (1971). On Group-theoretic Decision Problems and their Classification. Princeton Univ. Press.
Miller, G.A. (1900). On the product of two substitutions. Amer. J. Math. 22, 185-190.
Milnor, J. (1976). Hilbert's problem 18: on crystallographic groups, fundamental domains, and on sphere packing. In: Mathematical Developments Arising from Hilbert Problems, Part 2. Amer. Math. Soc.
Poincaré, H. (1882). Théorie des groupes fuchsiens. Acta Math. 1, 1-62.
Rotman, J.J. (1984). The Theory of Groups: An Introduction. 3rd edition, Allyn and Bacon.
Schwartzenberger, R.L.E. (1980). N-dimensional crystallography. Pitman.
Schwartzenberger, R.L.E. (1984). Colour symmetry, Bull. London Math. Soc. 16, 209-240.
Speiser, A. (1945). Die Theorie der Gruppen von Endlicher Ordnung. Dover. (Chapter 6: Symmetrien der Ornamente; Chapter 7: Die Kristallklassen).
Weyl, H. (1952). Symmetry. Princeton Univ. Press.
Wieting, T.W. (1982). The Mathematical Theory of Chromatic Plane Ornaments. Dekker.

INDEX